The Heat Will Kill You First

THE HEAT WILL KILL YOU FIRST

LIFE AND DEATH ON A SCORCHED PLANET

JEFF GOODELL

THORNDIKE PRESS
A part of Gale, a Cengage Company

Thorndike Press, a part of Gale, a Cengage Company.

Thorndike Press® Large Print Nonfiction.

The text of this Large Print edition is unabridged.

Other aspects of the book may vary from the original edition.

Set in 16 pt. Plantin.

LIBRARY OF CONGRESS CIP DATA ON FILE.
CATALOGUING IN PUBLICATION FOR THIS BOOK
IS AVAILABLE FROM THE LIBRARY OF CONGRESS.

ISBN-13: 979-8-88579-805-1 (hardcover alk. paper)

Published in 2024 by arrangement with Little, Brown and Company, a division of Hachette Book Group, Inc.

Print Number: 1   Print Year: 2024
Printed in Mexico

For Simone

Heat Index

30 million

Number of people who live in extreme heat today (above 85 degrees mean annual temperature)

2 billion

Number of people who are likely to live in extreme heat in 2070

1 mile per year

Average speed at which land animals are moving to higher, cooler latitudes

2.5 miles per year

Average speed at which malaria-carrying mosquitoes are moving to higher, cooler latitudes

210 million

Increase in number of people facing acute food insecurity since 2019

21%

Loss in global agricultural production in last 20 years due to climate-driven heat and drought

250,000

Annual worldwide deaths from firearms

489,000

Annual worldwide deaths from extreme heat

CONTENTS

Fuckin' heat . . . I . . . oh, man, I just . . . can't . . .
fuckin' . . . *make it*!

— Michael Herr, *Dispatches*

Prologue: The Goldilocks Zone

When heat comes, it's invisible. It doesn't bend tree branches or blow hair across your face to let you know it's arrived. The ground doesn't shake. It just surrounds you and works on you in ways that you can't anticipate or control. You sweat. Your heart races. You're thirsty. Your vision blurs. The sun feels like the barrel of a gun pointed at you. Plants look like they're crying. Birds vanish from the sky and take refuge in deep shade. Cars are untouchable. Colors fade. The air smells burned. You can imagine fire even before you see it.

In the summer of 2021, weathercasters in the Pacific Northwest warned people that a heat wave was on the way. Workers in Washington's Yakima Valley were summoned to cherry orchards at 1 a.m. so the ripe fruit could be picked before it turned to mush. Air-conditioning contractors were deluged with calls. Electric fans sold out at Home

Depot and Lowe's. The Red Cross activated its heat alert network, blasting out warnings to people to drink water and check on family and friends who lived alone. Libraries and churches set up cooling centers for the homeless or anyone who needed refuge. In Portland, Chris Voss, the emergency management director for Multnomah County, decided to open the Oregon Convention Center, which was capable of providing a cool retreat for hundreds of people. "What's coming is not just uncomfortable heat," Jennifer Vines, the lead health officer for the region, advised Voss. "This is life-threatening heat."

Nevertheless, the heat hit with a force that few people anticipated. The Pacific Northwest, after all, has long been seen as a climate refuge. It was the place you moved to if you wanted to live somewhere that was "safe" from climate change. There are beaches and lakes and stately trees and volcanic soil where anything grows, from blueberries to boxwood to grapes that are crushed into world-class Pinot Noir. There are glaciers in the Cascades and lush temperate rain forests in Olympic National Park and more than a few fragments left of the Edenic paradise that pulled so many settlers over the Oregon Trail. In the 1970s, Steve Jobs picked apples on a farm in the region and loved it so much he named a computer company after it. Heat wave? No big deal. This was not Phoenix, where heat owns the city. Or New Delhi, where heat is both a goddess and a demon. In the Pacific Northwest that summer, everyone might have known the heat was coming, but nobody thought it would be a searing, ghostly force that would melt asphalt and kill loved ones and force a new reckoning with the world they live in.

The heat wave had been born over the Pacific a week or so earlier. Atmospheric waves wobbled across the Northern Hemisphere, creating a high-pressure lid that allowed heat radiating up from the ocean to gather beneath it. As this pile of heat drifted to the coast, it grew quickly in size and intensity (land reflects and amplifies heat much more efficiently than water), creating what scientists called a heat dome. In a twenty-four-hour period, the temperature in downtown Portland jumped from 76 degrees to 114 degrees, the hottest temperature in 147 years of observations.* Suddenly, the ferny salamander-land of the Northwest felt like the hard-baked steel and sand of Dubai.

* Unless otherwise noted, all temperature references in this book are in Fahrenheit.

Ice, nature's most exquisite thermometer, registered the heat first. The last of the winter snow in the Cascades vanished from shady hollows in the forests and atop the glaciers near the peaks. With the protective snow-pack gone, the blue glacial ice itself began to melt, rushing down streambeds and canyons in a swirl of silty gray water, carrying ancient sediment from before the fossil fuel age, before books, before the pyramids. The rush of meltwater flooded roads and towns as it rolled down to the rivers and out to the sea. In the Columbia River, which is the largest river in the Northwest, the wash of sediment was so enormous that satellites circling the Earth photographed a gray plume flowing several miles into the Pacific.

In streams and rivers, migrating salmon immediately sensed the changes in water temperature. They had spent three or four years in the cold, salty Pacific and now were swimming upstream in freshwater back to where they were born to lay eggs and begin the cycle anew. The salmon's journey is one of the great wonders of nature. But it is also a fragile one. Warm runoff in the rivers — shallow water can heat up quickly as it flows down out of the mountains — made it difficult for the struggling salmon to breathe (the warmer the water, the faster oxygen molecules vibrate with kinetic energy, allowing them to flee their molecular bonds and escape into the air. "It leaves fish feeling like they are breathing with a plastic bag over their heads," one wildlife biologist told me). Their iridescent silvery skin broke out in red lesions. Cottony puffs of fungus grew on their backs. Some escaped to cooler tributaries. But tens of thousands of others, exhausted and suffocating and literally disintegrating in the warmth, became meals for other fish or washed up on the riverbanks, where they were picked apart by racoons and eagles.

In the mountains and valleys, every plant and tree was assaulted by the heat, rooted in place and unable to move, creators of shade that were themselves unable to seek refuge. As the temperature rose, they struggled with the heat just as humans do, trying to preserve water while the sun and the heat sucked it out of the soil and the flesh of their leaves and trunks. All across the Pacific Northwest there was a great clenching as the plants closed the pores on the undersides of their leaves, in effect holding their breath, hoping the heat would pass quickly. Blackberry and blueberry plants drank the moisture out of their own fruit, leaving it dry and withered on the stalk. On broadleaf trees like ash and maple, leaves brittled and curled. As temperatures rose, some of the most sun-exposed trees opened their pores, desperately trying to cool off by sweating. Their roots worked to pull water out of the dry soil, but instead sucked air bubbles into the veins that ran up their trunks, causing them to rupture. If you'd had the right kind of microphone, scientists say, you could have heard the trees screaming.

In the mountains, bighorn sheep headed for higher elevations. Doves rested on shady branches and opened their wings to aerate their bodies. And like dogs, they panted. Baby hawks, hot and fuzzy in their nests, faced the choice of overheating with their siblings or jumping out before they were ready to fly. Many jumped. Dozens of fluttering, broken bodies were found by hikers and taken to wildlife rehabilitation centers.

For some animals, however, these were good times. Caterpillars basked in the heat to kill pathogens in their bodies. Maggots hatched in the mouths of the dead salmon on the riverbanks. For pine bark beetles, an invasive species that is decimating western forests, the heat was like guzzling Red Bull. Their metabolism revved up, their appetite grew, and they moved like a marauding army through thousand-acre stands of Jeffrey pines.

In the cities and suburbs, air-conditioning units whirred. Overloaded power lines hummed and sagged. In the grid control centers, dispatchers sent urgent messages to power companies, who fired up idle natural gas plants, which can quickly generate electricity (and profits) during desperate times. In Oregon's Multnomah County, outdoor sporting events and concerts were canceled. Volunteers made thousands of calls to check in on disabled people and senior citizens. In Vancouver, British Columbia, police answered a surge of calls from people with difficulty breathing or in cardiac arrest. Sirens wailed and hospital emergency rooms were crowded with panting, red-faced people. Desperate to lower their body temperatures as quickly as possible, doctors filled body bags with ice and zipped them inside.

Vivek Shandas, a professor of urban studies and planning at Portland State University, drove around in his Prius with his eleven-year-old son, Suhail, measuring the temperature in different parts of the city. In Lents, one of Portland's poorest neighborhoods, where trees are few and concrete is plentiful, Shandas measured an air temperature of 124 degrees, the highest temperature he had ever recorded in fifteen years of chronicling heat. "When I stopped and opened my car door, the first thing I felt was my eyes burning," Shandas recalled. "My skin was on fire. It just feels like you're melting." He drove to Willamette Heights, a tree-lined suburb with parks and lots of greenery where the median house price is about $1 million. He measured the air temperature again: 99 degrees. In a heat wave, wealth can afford twenty-five degrees of coolness.

Nobody knows for sure how many people died during the seventy-two hours of extreme heat in the Pacific Northwest. The official count was 1,000, but heat is a subtle killer and doesn't always make it onto death certificates. The actual number is likely far higher. Whatever the true toll was, it certainly included Rosemary Anderson, sixty-seven, whose neighbor had texted her, "Good night, sleep tight, may you find lots of angel wings upon your pillow," the night before she was found dead in her house, where the indoor temperature measured 99.5 degrees. It also included Jollene Brown, sixty-three, who lived alone in an apartment a few miles away from Anderson, who was found by her son Shane sitting in her La-Z-Boy rocker with one foot on the footrest, one foot on the floor, as if she'd been about to stand up but couldn't because of the wall of heat in her tiny un-air-conditioned living room. As with most heat waves today, the people who died first were elderly folks who lived alone, or who were too poor to afford air-conditioning, or who had a medical problem that left them vulnerable. In this sense, a heat wave is a predatory event, one that culls out the most vulnerable people. But that will change. As heat waves become more intense and more common, they will become more democratic.

Even before this heat wave hit the Pacific Northwest, the forests were burning, dried to tinder by years of rising heat and declining rainfall. But in British Columbia, the heat brought something like spontaneous combustion to Lytton, an old mining camp (pop. 250) at the confluence of the Fraser and Thompson Rivers where First Nations people have lived for thousands of years. In the 1970s, the town was reborn as a whitewater rafting mecca due to its proximity to the spectacular flow of water through the black granite walls of Thompson Canyon. On the third day of the heat wave, temperatures in Lytton reached an unholy 121 degrees. Lorna Fandrich remembers looking out the window at the Chinese history museum in town — an institution she and her husband founded to commemorate and honor the Chinese workers who built the railroad and worked in the mining camps — and seeing leaves dropping off a tree, as if it were an autumn day, though it was only June. "I thought, *How strange,*" she said later. Then wind kicked up and a spark jumped out from the steel wheel of a passing freight train. Within minutes, the town was aflame. Mayor Jan Polderman raced his Honda minivan through the village, begging skeptical residents to flee. He picked up one of the last stragglers, a man running down a road with his cat in a cage.

Jeff Chapman, who lived with his parents outside of town, was just starting to cook dinner when he saw the smoke and fire approaching. "Ten minutes later, our house was fully engulfed," he said. "There was nothing we could do. We had nowhere to go." As the blaze swallowed the house and the trees around it, Chapman rushed his parents, who were both in their sixties, into a trench that had been dug a few days earlier to repair a septic system. It wasn't big enough for all three of them, so he grabbed a sheet of metal roofing nearby and laid it over them. Then he took refuge on some nearby railroad tracks, hoping the fire would pass.

That's when the power line crashed down across the trench where his parents were sheltering. "I knew my parents were in that hole and I'm watching the house burn and I'm thinking, *Oh my God.*" Chapman survived the inferno. His parents did not.

A few days later, as if by some miracle, it was blue skies and cool weather again in Lytton. The entire town was smoldering, burned to the ground. A few pieces of porcelain pottery survived in the ashes of the Chinese history museum. Douglas fir trees on the edge of town looked like black spears. There was grief and horror about what had happened, and vows to rebuild. Meanwhile, out on the coast, limp starfish and the shells of mussels and clams washed ashore by the millions. Chris Harley, a zoologist at the University of British Columbia, estimated that the three-day heat wave killed more than a billion sea creatures.

But as June came and went and summer turned to fall, life went back to normal, and the memory of the heat wave faded, as memories of heat waves always do, until they become like the fleeting images of a nightmare you're not quite sure you had. Or a future you don't want to imagine.

■■■■

You probably think of heat on a temperature scale, either Fahrenheit or Celsius. You think of it as a gradual, linear thing, a quality of the air around you that moves up and down in increments, or that can be controlled by a thermostat. Seventy degrees is a little hotter than 68 degrees, which is a little hotter than 65 degrees. The change of seasons also plays into this incremental perception of heat — winter gradually warms into spring, spring into summer. Yes, there are some days that are noticeably hotter or colder than others, but we crank up the air-conditioning or throw on a sweater. We trust it will pass and things will return to normal. Temperature is a merry-go-round that we are used to riding.

This sense of incrementalism also holds true with the climate crisis. The Earth is getting hotter due to the burning of fossil fuels. This is a simple truth, as clear as the moon in the night sky. So far, thanks to 250 years of hell-bent fuel consumption, which has filled the atmosphere with heat-trapping carbon dioxide (CO_2), global temperatures have risen by 2.2 degrees since the preindustrial era and are on track to warm by 6 degrees or more by the end of the century.

The more oil, gas, and coal we burn, the hotter it will get.

Right now we're more than halfway to 3.6 degrees (2 degrees Celsius) of warming from preindustrial temperatures, which scientists have long warned is the threshold for dangerous climate change. The reports of the UN's Intergovernmental Panel on Climate Change are full of harrowing details of what might happen to our world with 3.6 degrees of warming, from collapsing ice sheets to crop-killing drought. But to nonscientists — which is to say, most humans on the planet — 3.6 degrees of warming does not sound dangerous at all. Who can tell the difference between a 77-degree day and an 81-degree day? Then there are folks who argue that extreme cold kills people and causes all kinds of weather-related problems too so maybe a hotter world isn't such a bad thing after all.* Even the phrase "global warming" sounds gentle and soothing, as if the most notable impact of burning fossil fuels will be better beach weather.

The difficulty of understanding the consequences of heat is amplified by conventional notions of what it means to be hot.

* For more about extreme heat vs. extreme cold mortality, see note on p. 470–71.

In pop culture, hot is sexy. Hot is cool. Hot is new. Websites publish "hot lists" of the latest books, movies, TV shows, and actors. Facebook started in Mark Zuckerberg's Harvard dorm as a hot-or-not website called FaceMash, which ranked the attractiveness of Harvard women. Heat is an expression of passion — you feel hot for the guy at the bar or engage him in a heated debate. A person who is quick to anger is hot-blooded. Near the house where I live in Austin is a gym called Heat Bootcamp. Here, sweat is purifying, a sign of inner strength (a throwback to medieval times, perhaps, when heat was linked to masculinity through what philosopher Thomas Aquinas called "the elemental heat of the semen"). In Miami, one of the hottest cities in the US, where heat is a lethal threat to outdoor workers and where the city regularly floods due to rising seas caused by the melting ice sheets of Greenland and Antarctica, the basketball team is named, without irony, the Miami Heat.

In this book, my goal is to convince you to think about heat in a different way. The kind of heat I'm talking about here is not an incremental bump on the thermometer or the slow slide of spring into summer. It is heat as an active force, one that can bend railroad tracks and kill you before you even understand that your life is at risk. Scientists don't fully understand how fast this heat can move or where it will appear next (until it happened, a killer heat wave in the Pacific Northwest seemed about as likely as snow in the Sahara). But there is one thing scientists do know: this is a form of heat that has been unleashed upon us through the burning of fossil fuels. In this sense, extreme heat is an entirely human artifact, a legacy of human civilization as real as the Great Wall of China.

The amount of heat generated by our consumption of fossil fuels is difficult to grasp: by one measure, the ocean absorbs the equivalent of the heat released from three nuclear bombs every second. And because CO_2 stays in the atmosphere for thousands of years, it's not going to cool off when we finally stop emitting CO_2 into the atmosphere. All that will do is stop the *increase* in warming. It will not reverse it. Until we figure out a way to suck massive amounts of CO_2 out of the sky, we will be stuck with a hotter planet.

This heat we are pumping into the sky is the prime mover of the climate crisis. The climate impacts you hear about most often, from sea-level rise to drought to wildfires, are all second-order effects of a hotter planet. The first-order effect is heat. It is the engine of planetary chaos, the invisible force that melts the ice sheets that will flood coastal cities around the world. It dries out the soil and sucks the moisture out of trees until they are ready to ignite. It revs up the bugs that eat the crops and thaws the permafrost that contains bacteria from the last ice age. When the next pandemic hits, the chances are good it will be caused by a pathogen that leapt from an animal that was seeking out a cooler place to live.

As a force, heat is mysterious because its effects are both slow and fast. Think of parched wheatfields, slowly dried out by months of heat that pulls moisture out of the ground and lifts it into the sky. Then think of heat waves that are the cosmic equivalent of a bug zapper that kills you before you understand what's happening. Extreme heat penetrates every living cell and melts them like a Popsicle on a summer sidewalk. It reverses evolution, driving entropy and disorder. It is the widening gyre that the poet W. B. Yeats wrote about,* an extinction force

* The opening lines of Yeats' "The Second Coming," written in 1919, soon after the end of World War I: "Turning and turning in the widening gyre/

that takes the universe back to its messy beginnings. Before there was light, there was heat. It is the origin of all things and the end of all things.

You don't need to be a Hollywood screenwriter to imagine how our world will be changed by extreme heat. A few things are self-evident.

As the temperature rises, it will drive a great migration — of humans, of animals, of plants, of jobs, of wealth, of diseases. They will all seek out cooler ecological niches where they can thrive. Some will fare better than others. Robins can migrate more easily than elephants. Poison ivy can move more quickly than an oak tree. Farmers who grow wheat have more options than farmers who grow peaches. And some creatures have nowhere to go. Polar bears in the Arctic can't migrate farther north. Frogs in Costa Rica aren't going to hop up to Canada.

Humans are better off than many plants and animals. With the help of technology, we can adjust to a lot of things. As one architect told me, "If you have enough money, you can engineer your way out of anything."

The falcon cannot hear the falconer/Things fall apart; the centre cannot hold"

And in some ways, he's right. If we can send photos through the air and drive a rover around on Mars, we can design new ways to live in hot places. You can see it happening right now in Paris and Los Angeles and many other cities around the world, where shade trees are being planted and streets painted white to deflect sunlight. Plant geneticists are developing new strains of corn and wheat and soybeans that can better tolerate high temperatures. Air-conditioning is becoming cheaper and more widely used. Communication from public health officials about how to protect yourself during a heat wave is improving. Clothing companies are developing new high-tech fabrics to reflect away sunlight and dissipate heat more quickly.

But even for the wealthy and privileged, adaptation to extreme heat has its limits. And the notion that eight billion people are going to thrive on a hotter planet by simply cranking up the air-conditioning or seeking refuge under a pine tree is a profound misunderstanding of the future we are creating for ourselves. In western Pakistan, where only the richest of the rich have air-conditioning, it's already too hot for humans several weeks a year. Planting a few thousand trees is not going to save them. In India, I talked with families who live in concrete slums that are so hot they burn their hands opening doors. Holy cities like Mecca and Jerusalem, where millions gather on religious pilgrimages, are caldrons of sweat. In the summer of 2022, nine hundred million people in China — 63 percent of the nation's population — suffered under a two-month-long extreme heat wave that killed crops and sparked wildfires. "There is nothing in world climatic history which is even minimally comparable to what is happening in China," one weather historian declared.

In a world of heat-driven chaos, heat exposes deep fissures of inequity and injustice. Poverty equals vulnerability. If you have money, you can turn up the air-conditioning, stock up on food and bottled water, and install a backup generator in case there's a blackout. If things get bad enough, you can sell your house and move to a cooler place. If you're poor, on the other hand, you swelter in an uninsulated apartment or trailer with no air-conditioning or an old, inefficient machine that you can't afford to run. You can't move somewhere cooler because you're afraid of losing your job and you don't have the savings to start over. "We're all in the storm, but we're not in the same boat," Heather McTeer Toney, the former mayor of Greenville, Mississippi, said during testimony before the US Congress. "Some of us are sitting on aircraft carriers while others are just bobbing along on a floatie."

Researchers at Dartmouth College estimate that since the 1990s, extreme heat waves amped up by climate change have cost the global economy $16 trillion. Heat lowers children's test scores and raises the risk of miscarriage in pregnant women. Prolonged exposure increases death rates from heart and kidney disease. When people are stressed by heat, they are more impulsive and prone to conflict. Racial slurs and hate speech in social media spike. Suicides rise. Gun violence increases. There are more rapes and more violent crime. In Africa and the Middle East, studies have found a link between higher temperatures and the outbreak of civil war.

The harshest truth about life on a superheated planet is this: as temperatures rise, a lot of living things will die, and that may include people you know and love. A study in *The Lancet,* a prestigious medical journal, estimated that 489,000 people worldwide died from extreme heat in 2019. That's far more than all other natural disasters combined, including hurricanes and wildfires. It is also more than the number of deaths from guns or illegal drugs. And those are only the deaths that are directly attributable to heat. There are also deaths caused by the heat-related amplification of ground-level ozone pollution (aka smog), or the smoke from wildfires in desiccated forests. The smoke can drift thousands of miles, lofting tiny particulates into the atmosphere. When you inhale them, they can trigger a variety of health problems, from asthma to heart attacks. The toll is enormous: globally, between 260,000 and 600,000 people die each year inhaling smoke from wildfires. Smoke pollution doesn't only kill people near fires either. Wildfires in western Canada have been directly linked to spikes in hospitalizations three thousand miles away on the East Coast of the US.

Earth's history is full of wild temperature swings, driven by volcanic eruptions, meteor strikes, and geologic mayhem. There have been palm trees in the Arctic and two thousand feet of ice over New York City. But for the last three million years or so, while humans evolved, the climate has been relatively stable. Stable enough, anyway, that our ancestors could migrate, adapt, and thrive.

But those days may be over. The last time the Earth was hotter than it is today was at least 125,000 years ago, long before anything that resembled human civilization appeared. Since 1970, the Earth's temperature has spiked faster than in any comparable forty-year period in recorded history. The eight years between 2015 and 2022 were the hottest on record. In 2022, 850 million people lived in regions that experienced all-time high temperatures. Globally, killer heat waves are becoming longer, hotter, and more frequent. One recent study found that a heat wave like the one that cooked the Pacific Northwest is 150 times more likely today than it was before we began loading the atmosphere with CO_2 at the beginning of the industrial age. The ocean, which hundreds of millions of people depend on for their food supply and which has a big influence on weather, was the hottest ever recorded in 2022. Even Antarctica, the coldest place on Earth, is not immune. In March of 2022, a heat wave invaded the ice-bound continent, pushing temperatures seventy degrees — *seventy degrees!* — above normal.

Extreme heat is remaking our planet into one in which large swaths may become inhospitable to human life. One recent study projected that over the next fifty years, one to three billion people will be left outside the climate conditions that gave rise to civilization over the last six thousand years. Even if we transition fairly quickly to clean energy, half of the world's human population will be exposed to life-threatening combinations of heat and humidity by 2100. Temperatures in parts of the world could rise so high that just stepping outside for a few hours, another study warned, "will result in death even for the fittest of humans."

Life on Earth is like a finely calibrated machine, one that has been built by evolution to work very well within its design parameters. Heat breaks that machine in a fundamental way, disrupting how cells function, how proteins unfold, how molecules move. Yes, some organisms can thrive in higher temperatures than others. Roadrunners do better than blue jays. Silver Saharan ants run across superhot desert sands that would kill other insects instantly. Microbes live in 170-degree hot springs in Yellowstone National Park. A thirty-year-old triathlete can handle a 110-degree day better than a seventy-year-old man with heart disease. And yes, we humans are remarkable creatures with a tremendous capacity to adapt and adjust to a rapidly changing world.

But extreme heat is a force beyond anything we have reckoned with before. It may be a human creation, but it is godlike in its power and prophecy. Because all living things share one simple fate: if the temperature they're used to — what scientists sometimes call their Goldilocks Zone — rises too far, too fast, they die.

1. A Cautionary Tale

When the babysitter arrived to take care of Miju at around 11 a.m. on Monday, August 16, 2021, she was surprised to find the house empty. Miju was the one-year-old daughter of Jonathan Gerrish and Ellen Chung, who had recently fled the Bay Area to start a new life in California's Sierra Nevada foothills, not far from the old Gold Rush town of Mariposa. Their modern three-bedroom house sat on ten acres of lightly forested land. It had wood floors and a big stone chimney and tall rectangular windows that looked over a rugged treeless canyon called Devil's Gulch. From the second-floor bedroom, you could just see the top of El Capitan, the iconic granite formation in Yosemite Valley, about thirty-five miles to the east. The house was their refuge from the hustle of Silicon Valley, where Gerrish worked as a software engineer at Snapchat, the instant messaging app company.

The babysitter, who had a key, let herself in and called out their names. No response. It had been a hot weekend, but the inside of the house was cool, thanks to the air conditioner, which was going strong. Oddly, Chung and Gerrish had left behind their wallets. Even more confusing, the diaper bag that the couple always took with them was there.

The babysitter had last seen them the previous Friday when she had finished straightening up the house. That evening, Chung had happily texted her a video of Miju starting to walk. She had made no mention of plans to be away that Monday. Gerrish and Chung, who doted on Miju and seemed deliriously happy with their new life in the Sierra foothills, were not the kind of people who would disappear for a last-minute road trip to Las Vegas.

Worried, the babysitter called the construction manager who was working with Gerrish on another home he owned, and whom she knew was in frequent contact with him. The construction manager was not initially concerned because Chung and Gerrish were a "very active family," an investigator later wrote in the police report. Nevertheless, the babysitter and the construction manager started making calls and sending texts to friends, asking if anyone had seen the couple. Steve Jeffe, a friend who lived in Mariposa, posted on Facebook: "Hi, please has anyone been in contact with Jonny Gerrish or Ellen Chung in the last two days . . . Please." By 5 p.m. that day, several friends began driving around looking for the family. At 11 p.m., they called the Mariposa County sheriff.

A few hours later, a deputy sheriff found Gerrish's truck at the Hites Cove trailhead a few miles from their home. By 4 a.m., a search and rescue team was on the scene. They headed down the trail in an ATV, flashlights cutting through the darkness. They radioed back that they had found tracks on the trail. But when they followed the tracks down to the Merced River, the tracks disappeared. By that time, the sun was rising. A helicopter was called in. More search and rescue teams arrived. One team headed down a steep trail with numerous switchbacks toward the river. They were a mile and a half down the trail when, at about 9:30 a.m., they discovered the bodies of Gerrish, Miju, and their dog, Oski. Gerrish was in a seated position, with Miju and Oski beside him.

At first, the search and rescue team saw no sign of Chung. About a half hour later, a deputy walking back up the trail from where Gerrish was found noticed "some disturbed dirt on the uphill side of the trail that appeared that something or someone had tried to go up the hill." He spotted a shoe, then Chung's body. Investigators would later conclude that the family had been hiking up the trail when they died. The location of Chung's body indicated that she had abandoned the trail and was climbing straight up the mountainside — a sign, investigators believed, of the urgency of their situation and her desperate attempt to reach their truck.

But even if Chung had made it to the truck, she might not have been able to get in. During a search of the area, investigators discovered a Ford key fob on the trail about a hundred feet below Gerrish's body. Had it accidentally fallen out of his pocket? Or did he have it in his hand and drop it and not realize it — perhaps a sign that he had been panicked or disoriented?

Rescuers found no signs of foul play. No marks on the bodies, no obvious signs of struggle. Because of the remote location and the difficulty of the terrain, the bodies could not be removed immediately. Instead, two deputies spent the night at the scene, guarding the bodies from bears and coyotes. The next morning, a California Highway Patrol helicopter airlifted the family off the trail.

Gerrish and Chung had moved to Mariposa about a year and a half earlier, just before Miju, their first child, was born. They had been living in San Francisco, where Chung taught yoga while finishing her degree in counseling psychology, and Gerrish wrangled computer code at Snapchat. But then Miju came along and the Covid-19 pandemic erupted and they needed a change. They decided they wanted to get out of the city and raise Miju closer to nature. Mariposa, which is just an hour outside the entrance to Yosemite National Park, was an ideal mix of wilderness and charm. "They fell in love with the Mariposa area," one friend of the family said.

Gerrish was born in Grimsby, an old fishing port in the northeast of England, where his father was an elementary school teacher and his mother a receptionist in a doctor's office. His brother Richard, who was two years younger, recalled their mother and father dragging them on long walks when they were kids. "My brother and I built dams across mountain streams and played manhunt (a more exciting version of hide and seek) in the woods," Richard wrote in an essay about his childhood. "But usually we ended up complaining about the distance we walked and crying because I was tired, hungry and my feet were sore." After high school, Gerrish attended Newcastle University, majoring in computer science. He worked for a few software companies in England, then got a job with Google in London. When the company offered him the chance to move to California, he jumped at it. "Jonny was somewhat awkward in his own skin growing up," Richard told me. "When he moved to San Francisco, he found his people. He loved it there."

Gerrish was six foot four and a bit shaggy, with a beard and longish hair that looked like a comb had never run through it. He wore Vans and supported Greenpeace and listened to techno and deep house music. Burning Man, the clothing-optional psychedelic super-rave that is held in the Nevada desert during the late summer, was his most holy holiday. His friends (of whom there were many) called him Jonny, a name that captured his boyish enthusiasm and charm. "You don't see many men happier than Jonny," a friend told me.

Chung grew up in Orange, California, and graduated from University of California, Berkeley in 2012. Her parents had immigrated from South Korea in the 1970s and eventually opened a successful restaurant in Orange. After graduation, she worked for a few years in marketing at a tech company but decided she wanted a change. She enrolled in classes at the California Institute of Integral Studies, a private university in San Francisco with roots in Eastern culture and philosophy, where she discovered she was drawn to — and good at — helping people talk about their problems. She wore stylish straw hats and loved the way the light filtered through the redwoods on the California coast and the long, open vistas at Zion National Park in Utah.

Gerrish and Chung both doted on Miju. You can see it in every photo of them together, the big beaming happy-father smile on Gerrish's face, the joy and new-mother exhaustion on Chung's. And Miju, waking up to the world, her eyes wide. She was just beginning to walk, to track birds flying across the sky, to take in all the wonders around her. Gerrish and Chung were protective of their young daughter, and careful about her surroundings. At one point, they asked a local contractor to make their daughter's bedroom cooler because it was "too stuffy."

The day before their hike, Gerrish mapped

out the route using the AllTrails app on his Google Pixel 4. The app helps users find local trails, giving maps and elevations, as well as a place for other hikers to leave comments. Gerrish had logged sixteen hikes in 2021, most of them three or four miles long, all of them in the mountains and canyons near his house.

The hike he planned for his family was not a remote wilderness adventure. It started at a trailhead a few minutes' drive from their house and ended at the top of Devil's Gulch, which was practically in their front yard. The trail went along a ridge, then down fairly steeply to the south fork of the Merced River, which flowed out of Yosemite and through the canyons toward Mariposa. The trail wandered along the flat ground on the banks of the river for about three miles. From there, Gerrish plotted a right turn, which would take them on a steep 2,300-foot climb through Devil's Gulch and back to their truck. All in all, it was an eight-mile loop.

Gerrish loved nature, but he was not a serious outdoorsman. His brother Richard, who now lives in Scotland with his wife and four children, had spent years as an Outward Bound leader, guiding teenagers on adventures into the wilderness. Richard had also rappelled into some of the deepest caves in the world (including one in Austria called Fit for Insane Worms and Geckos). Gerrish, in contrast, was more of a weekend adventurer. The construction manager who worked with him renovating one of his houses called Gerrish and Chung "city folk," pointing out that Gerrish would go to the store and get firewood rather than chopping his own.

As it happened, Gerrish called Richard the day before the hike for some parenting advice. Gerrish mentioned to his brother that he had been out exploring the property that day and that it had been unusually hot. Gerrish also mentioned that he was planning a family hike the next day to scout out possible swimming holes on the Merced River. Richard, who was well aware of the dangers of hiking in the heat, cautioned his brother, telling him to carry plenty of water and get an early start. Gerrish agreed, and promised they would be off the trail before it got too hot.

On Sunday morning, Gerrish and Chung were up at dawn. They skipped breakfast and gathered up their gear: hiking poles, baby carrier, diapers and sippy cup for Miju, and a leash for eight-year-old Oski, a big strong dog that was part Akita. For drinking water, Chung carried an Osprey hydration pack, which held 85 ounces (about two and a half quarts) of water. Gerrish wore dark shorts, a yellow T-shirt, and tennis shoes. Chung wore hiking boots, spandex shorts, and a yellow tank top. They woke up Miju and dressed her in a short-sleeved onesie and pink shoes. Then they loaded everything into their 2020 gray Ford Raptor, an off-road version of the F-150 pickup, and headed out for the five-minute drive to the trailhead.

At about 7:30 a.m., a woman walking her dogs along Hites Cove Road, which was not much more than a narrow dirt track, saw their truck drive past and park at the trailhead. Gerrish took their first selfie of the family on the trail at 7:44 a.m. The temperature was in the midseventies, not humid, a warm but lovely morning. Under normal conditions, Gerrish might well have calculated that the eight-mile loop might take four or five hours to complete. If all went well, they would be off the trail by 1 p.m., just as the afternoon sun began to blaze.

The Sierra foothills are still marked by the California Gold Rush of the 1850s and 1860s. You see piles of old tailings along the rivers, abandoned mining shacks and sluices. In the Mariposa area, quartz veins — the geological homeland of gold — run twelve feet thick through the mountains. Hites Cove, where Gerrish and Chung were hiking, was once a mining camp with over a hundred people. Gold Rush fever is long gone, but you still occasionally run into prospectors with metal detectors wandering through the area. These days, most hikers come for the spring wildflowers, especially the spectacular fields of orange California poppies, which thrive in hot, dry, rocky soil. While hiking, you might see a bear, bobcat, or coyote (or, more likely, their scat). At the bottom of the canyon, big rainbow trout rest in the deep pools and eddies of the Merced.

In recent years, climate-driven heat and drought have turned the area into a tinderbox. The area was badly burned in the 2018 Ferguson Fire, which consumed almost a hundred thousand acres and forced Yosemite National Park to close for the first time in decades. Two firefighters died. The fire, caused by a spark from the catalytic converter of a vehicle, turned the dry grass and bark-beetle-infested trees into an inferno. In the three years since the fires, wildflowers had returned and a few saplings were rising out of the rocky soil, but most of the trees were charred sticks poking up at the sky and offering little shade for hikers or wildlife on a hot afternoon.

Even before the 2018 fire, the trail that Gerrish had selected for the hike was a risky choice. The steep ascent out of the canyon was along a southeastern-facing slope, which meant it was exposed to the full brutality of the sun. "It's a horrible trail," one local wrote on social media. "With the poison oak, rattlesnakes, and potential for broken ankles, it just isn't worth it." Another local, who hiked the trail on a mild spring day, praised the wildflowers blooming on the mountainside, but noted that it was dangerously exposed: "I wouldn't want to do this [hike] on a hot day."

For Gerrish and his family, the hike started off easy. The first two miles were mostly downhill. The morning sun would have felt good, the light slanting across the mountains. It only took them a little over an hour to get to the river, where they stopped to take another family selfie at 9:05 a.m. They rambled along the river for the next hour and a half, where they may well have stopped for a drink from their hydration pack and even doused their hands and faces in the cool river water.

At 10:29 a.m., they took a final family selfie along the river, then began the climb.

It had been three hours since they left the truck. The temperature had risen to nearly 100 degrees and it was getting hotter with each passing minute. The recently burned trees along the steep trail were black and leafless. The tall grass was sunburned to a golden, crispy brown, like straw.

If there is one idea in this book that might save your life, it is this: The human body, like all living things, is a heat machine. Just being alive generates heat. But if your body gets too hot too fast — it doesn't matter if that heat comes from the outside on a hot day or the inside from a raging fever — you are in big trouble.

Every organism manages heat in a different way (more about that in an upcoming chapter). We humans work hard to maintain an internal temperature of about 98 degrees, no matter what the temperature is outside. If it is cold outside, we pull blood into our vital organs to keep them warm. If it is hot outside, we push blood toward our skin so it can be cooled by sweat. That's why dry heat is often less dangerous than wet heat — the more humid the air is, the more difficult it is for our sweat to evaporate and dissipate heat. And like all living things, our bodies have thermal limits. Those limits vary depending on age, general health, and a number of other factors. But there is general consensus among researchers that a wet bulb temperature of 95 degrees — which basically means both outdoor air temperature and humidity levels are high (see glossary for a more precise definition of wet globe temperature) — is the upper end of human adaptability to humid heat. Beyond that, our body generates heat faster than it can dissipate it.

And that's where trouble begins. Hyperthermia, or abnormally high body temperature, causes a range of physiological responses that might start with dizziness and heat cramps and lead all the way to heatstroke — a condition that can be, and often is, fatal.

Broadly speaking, there are two kinds of heatstroke: classic and exertional. Classic heatstroke hits the very young, the elderly, the overweight, and people suffering from chronic conditions like diabetes, hypertension, and cardiovascular disease. Alcohol and certain medications (diuretics, tricyclic antidepressants, antipsychotics) can increase one's susceptibility. Classic heatstroke is often what happens to babies who are left in cars, or older people trapped in upper-story apartments with no air-conditioning on summer days.

Exertional heatstroke, on the other hand, often hits the young and fit. Exercise drastically accelerates temperature rise. Anytime you flex a muscle, it generates heat. In fact, when you move a muscle, only about 20 percent of the energy you expend actually goes to muscle contraction; the other 80 percent is released as heat. That is why marathon runners, cyclists, and other athletes sometimes push into what is called exercise-induced hyperthermia, where internal body temperatures typically hit 100 to 104 degrees. Usually, there's no lasting damage. But as your body temperature climbs higher, it can trigger a cascading disaster of events as your metabolism, like a runaway nuclear reactor, races so fast that your body can't cool itself down.

It doesn't take long. And being young or in great shape won't save you. In fact, being young and strong allows you to fight off the warning signs of heat exhaustion until it is too late. A few years ago, eighteen-year-old Kelly Watt, a track star and aspiring journalist in Virginia, parked his car on a hilly road where he often trained and went out for a fifty-minute run on a hot summer afternoon. A few hours later, Watt's father found his body in some bushes not far from his car. Handprints on the car suggested Watt had made it back to the car after his run, but had become disoriented from heat, couldn't open the door, and then wandered off into the bushes, where he collapsed and died.* In 2021, Philip Kreycik, a thirty-seven-year-old ultramarathoner and father of two young children, drove up into the hills near Pleasanton, California, on a July morning to go for a run. He parked his Prius on a dirt road, left his water bottle in the center console, and headed out. By noon, the temperature was 105 degrees. After Kreycik's wife reported him missing a few hours later, hundreds of people rallied for a search and rescue effort that the Alameda County Sheriff's Office called one of the largest ever on the West Coast. Twelve thousand people joined a Facebook group to help with the search and a fundraiser for Kreycik's family has raised more than $150,000. His body was finally found in a remote area twenty-four days later. Cause of death: hyperthermia.

* Watt wrote a weekly sports column, "Sports Wrap," in a Charlottesville paper called *The Hook*. It's a tragic irony that on the day he died, he overslept because he had been up late the previous night, writing about the heatstroke death of jockey Emanuel Jose Sanchez at Colonial Downs, a racetrack near Richmond.

Both Watt and Kreycik were excellent athletes. Both knew it was going to be hot during their run. Neither carried water. But would it have mattered? In 2016, thirty-four-year-old Michael Popov, one of the world's top ultrarunners, who routinely ran hundreds of miles in the rugged Sierra Nevada mountains, set out on a six-mile jog in Death Valley on a hot August day. He was carrying four bottles of water and ice. Two hours later, he was found collapsed on the side of a road. He died later that day.

There is a lot of confusion about the relationship between water and heat exhaustion and heatstroke. Water is necessary to keep sweat flowing. If you get dehydrated, you can't sweat. And if you can't sweat, you can't cool off. But drinking water does not in itself cool off inner-core body temperatures. Put another way, dehydration can exacerbate heat exhaustion and heatstroke, but you can still die of heatstroke and be well hydrated. In one study in Montana, a wildfire fighter working in extreme heat for a seven-hour period who continuously drank a huge amount of water — more than twice as much as the firefighters around him — still had a core body temperature of 105 degrees, which is well into heatstroke-land.

Sam Cheuvront, a heat and hydration

expert who spent more than twenty years at the US Army Research Institute of Environmental Medicine in Massachusetts, put it to me this way: "Both heat exhaustion and heatstroke can occur in the absence of dehydration. We can speculate that proper hydration can, however, delay heat exhaustion because dehydration exacerbates heat exhaustion. But proper hydration cannot prevent heatstroke."

Drinking water when it's hot is certainly important. A common recommendation is about one half quart (16 ounces) of water per person per hour of moderate activity in moderate temperatures. But in extreme conditions, even that isn't enough. A well-hydrated human can sweat up to three quarts per hour, but no matter how much water you drink, your body can only replace about two quarts of water per hour — so if you are in a hot place for a long time, dehydration is a concern.

Even at a sweating rate of two quarts per hour, which is what a firefighter might do while working in a hot environment and wearing protective clothing, it takes an hour to exceed 2 percent dehydration, which is really where dehydration starts contributing to heart strain, mostly due to reduced blood volume. It also exacerbates the competition for blood flow between your muscle, your skin, your brain, and your organs.

The only effective treatment for heatstroke is to get a person's core body temperature down, fast. A cold shower or bath, or tubs (or, as I mentioned in the prologue, body bags) of ice, is one way to do this. Another is to quickly cool places on the body where, because of the structure of our veins, a lot of blood circulates close to the surface: the bottoms of the feet, the palms of the hands, the upper part of the face. Taking Tylenol or aspirin doesn't help. In fact, both can interfere with kidney function, and make it harder for your body to deal with rising temperatures. Only after core body temperature is lowered do the damages from heatstroke stop and hope of repair and recovery begin.

About an hour and a half after Gerrish, Chung, Miju, and Oski left the banks of the Merced, they were in trouble. They had climbed two miles up the trail, but still had another mile and a half of steep switchbacks to go before they got back to their truck.

At 11:56 a.m., Gerrish pulled his phone out of his pocket and attempted to send a text: "[name redacted] can you help us. On savage lundy trail heading back to Hites cove trail. No water or ver [over] heating with baby." Records would later show that the air temperature at that time was 107 degrees. But on the trail, with full sun and no shade and rocks absorbing and amplifying the heat, the actual temperature that Gerrish and his family were experiencing was certainly much higher.

Gerrish and Chung surely had a moment when they stopped to consider whether it was better to stop climbing and turn around and seek refuge by the river. There wasn't a lot of shade down there either, but there was some. And they could have found some relief by wading out into the cool water. But at the same time, if they retreated to the river, they would still have to hike out eventually, and the afternoon was just going to get hotter. Waiting until the temperature dropped and the sun lost its edge meant waiting until late afternoon or early evening. While that might have been the safe decision, it had its own risks: they were out of water, and signs along the river warned hikers not to drink from it due to toxic algae. The actual risk of getting sick from toxic algae was extremely low compared to the danger of heatstroke, but Gerrish and Chung may not have known that.

There was also the issue of food for Miju. They had not brought enough diapers or baby formula with them for an entire day. Perhaps, out of love for Miju, they decided that it was better to suffer in the heat themselves, push through the climb back to the truck, blast the AC, and feel the relief of having escaped a heat-driven nightmare.

The typo in Gerrish's text message ("or ver") may be nothing more than a sign that he was in a hurry to get the message out. But it also could be a sign that the heat was already causing him some cognitive difficulty, which is common during extreme heat exhaustion. If that was the case, it would make clear-headed decision-making about whether to keep climbing in the heat or seek refuge near the river all the more difficult.

Whatever he was thinking at that moment, Gerrish was clearly aware that their situation was growing more dire. Over the next twenty-seven minutes, he attempted five phone calls, but because of the lack of service in the area, none went through. He did not call 911. If he had, there's a chance he might have gotten a response, given that in remote areas, 911 calls are routed differently to cell towers and so are sometimes picked up in areas where other calls are not. Gerrish may not have been aware of this, or he simply might have been too disoriented to consider it. In any case, he made one last attempt to call someone for help at 12:36 p.m. By then, two hours had passed since they left the shaded banks of the Merced.

A few years ago, on a hot day in May, I climbed Maderas volcano in Nicaragua. The volcano, which is on Ometepe Island in Lake Nicaragua, is a popular hike for travelers. The trail winds through a dense rain forest where brightly colored parrots flash by and troops of spider monkeys loll in the trees. The climb to the top of the volcano was six miles round trip, with an elevation gain of 3,750 feet. That was a steep climb, but I was in good shape, healthy, no medical issues. So why not? Early one morning, I took a bus from a nearby village where I was staying to Lake Nicaragua. At a little shack on the beach, I met up with a local guide I'd hired to accompany me on the hike (required by law in Nicaragua). I could see the volcano in the distance. It was stark and symmetrical, like something a six-year-old would draw.

Before we started, my guide, Roberto, made sure we had plenty of water with us. We also had some nuts and candy bars and dried fruit. Roberto didn't speak English and I didn't speak much Spanish. He was wearing an Ohio Buckeyes T-shirt and carrying a small pack.

It was warm when we started, maybe 80 degrees, and humid. That was already plenty warm for me, given that it was early fall and I was living in upstate New York at the time, where the weather was crisp and cool. I had only been in Nicaragua a few days, so I'd had little time to get acclimatized to the heat. I have since learned that after you spend a few weeks in a hot climate, your body makes subtle adjustments that help you better tolerate heat stress.* Your normal deep body temperature drops. Your body sweats at a lower temperature, and so there is less strain on your heart, which keeps your heart rate from rising fast. At the same time, your heart pumps more blood per stroke. Your body retains more fluids and blood volume rises, increasing water reserves for sweating and cooling. But these changes are not permanent. "If you go out of

* Exactly how long it takes to acclimatize to a hotter climate depends on a variety of factors, including how much you exercise in the heat and your physical condition. Two weeks is average for most people. Interestingly, exposure to air-conditioning even for brief periods slows or even stops acclimatization.

the heat, within a few weeks all the adaptations go back to zero," says Sam Cheuvront.

I had not given my body time to do any of that. In fact, I had not given the dangers of climbing a volcano on a hot humid day any thought at all. The idea that I might have heatstroke seemed about as likely to me as being abducted by aliens.

The trail was muddy and steep, cutting through the rain forest. We climbed at a slow, steady pace. My legs felt fine. I listened for monkeys. I wondered if there were jaguars. I tried to make simple conversation with Roberto, who was friendly enough but clearly preferred to hike in silence. He turned around frequently to make sure I was okay.

At some point, maybe an hour into the hike, I noticed I was sweating a lot. That was not surprising. It was hot and humid, and I was climbing a steep volcano. We stopped for a few minutes so I could catch my breath. I noticed Roberto wasn't sweating as much as I was, and I thought, *Well, maybe I'm not in as good shape as I think I am.* But I wasn't that tired, my legs were fine. I drank some water.

I climbed for another twenty minutes or so, still sweating, still hot, but feeling okay. Then a strange thing happened. I crossed over a threshold. I started sweating uncontrollably, just dumping water out of my pores. My heart pounded, blood rushed to my face. At the same time, my skin felt cold and clammy. It was like I was burning up and chilling out simultaneously.

I sat down on a log in the shade. *"Caliente,"* I said. Roberto looked at me concerned, and said something in Spanish that I didn't understand but that I took to mean, "You need to rest and drink water, you fool." The sweating got worse. My shirt was soaked. My pants were soaked. My heart raced faster. I had no idea what was happening to me. I was dizzy, ready to faint. My heart felt like it was going to explode.

Let me be clear: My situation was very different from what the Gerrish family faced in California. For one thing, I was in a rain forest, which meant I wasn't spending a lot of time with direct sunlight beating down on me. Also, the air temperature was lower. I don't know precisely how hot it was during my climb, but I'm guessing the temperature had risen to 90 degrees by then, which is 20 to 25 degrees cooler than what the Gerrishes faced. On the other hand, it was far more humid during my hike, which meant the air had very little capacity to absorb all the sweat that was rolling off my body. And with no evaporation, there was little chance of cooling.

But there was one similarity: we were equally ignorant about the risks of heat. Everyone understands the dangers of riding a bike without a helmet or going for a drive in a car without a seatbelt. Or the risks of smoking, which I understood personally after my father died of lung cancer at the age of fifty-three. But heat? To me, heat was just temperature. It wasn't a lethal weapon.

I hovered in this strange state for ten, fifteen minutes. I remember looking at the sweat pouring off me, wondering how my body could contain that much water. Luckily, we had plenty of water with us. I drank and drank. It seemed to pour right through me, the water passing through the membranes of my cells as if they were leaky water balloons. I was afraid I was going to die right there in the rain forest, in the mud, with poor Roberto watching me but unable to do anything.

Then the sweat receded. My heart slowed. I felt my body calming down and reasserting control. None of this was the result of anything I did consciously or deliberately. I drank more water. After another five minutes or so, the sweating stopped. I was soaked, but I felt stable, if a little weak. I ate some dried fruit that I had stashed in my backpack. Eventually I stood up, smiled at Roberto, and told him I was ready to go.

We continued up the volcano. Roberto stuck closer to me, as if he were afraid I might collapse at any moment. An hour or so later, we reached the summit, with a spectacular view into the lagoon in the crater. We sat and ate lunch. By that time, I had drunk all the water we had carried with us — about four quarts. I knew the rest of the hike was all downhill, so presumed I would be okay.

And I was. We hiked down the volcano without incident. That evening, I drank several cold *cervezas* at a little bar in the rain forest. I felt like I was trying to fill a dry hole inside me. But until I began writing this book, I had no real understanding of how much trouble I had been in, and how lucky I was.

No one knows the exact sequence of events that led to tragedy for the Gerrish family. But on a hot day, the road to heatstroke looks like this: As soon as you step outside, your blood grows warmer, heated by the sun's radiation and your own rising metabolism. Keeping your core body temperature at around 98 degrees — the happy place for humans — now requires work. Receptors in the preoptic area of your brain's hypothalamus start to fire, telling your circulatory system to push more blood toward your skin, where the heat can be dissipated. Your sweat glands begin to pump salty liquid from a tiny reservoir at the base of the gland up to the surface of your skin. You sweat. As the sweat evaporates, the heat is carried away.

But the amount of heat your body can dissipate through sweat is limited. Your blood vessels dilate, trying to move as much overheated blood to the surface as possible. But if you don't find a place to cool off, your core body temperature will rise quickly. And the harder you are working your muscles, the faster it will rise. Your heart pumps madly, trying to push as much blood as possible to your skin to cool off, but it can't keep up. As your blood is shunted away from your core, your internal organs — your liver, your kidneys, your brain — become starved for blood and the oxygen it carries. You feel light-headed. Your vision dims and narrows. As your core body temperature rises to 101 degrees, 102 degrees, 103 degrees, you feel wobbly — and due to the falling blood pressure in your brain, you will likely pass out. This is in fact an involuntary survival mechanism, a way for your brain to get your body horizontal and get some blood to your head.

At this point, if you get help and can cool down quickly, you can recover with little permanent damage.

But if you fall onto the ground in a place that is exposed to the sun and lie there, the dangers increase. It's like falling into a hot frying pan. Ground temperatures can be twenty to thirty degrees above air temperature. Your heart becomes desperate to circulate blood and find a way to cool down. But the faster your heart beats, the more your metabolism increases, which generates more heat, which causes your heart to beat even faster. It's a lethal feedback loop: As your internal temperature rises, rather than cranking up your air conditioner, your body fires up your furnace. If you have a weak heart, that might be the end for you.

At a body temperature of 105 to 106 degrees, your limbs are convulsed by seizures. At 107 and above, your cells themselves literally begin to break down or "denature." Cell membranes — the thin lipid walls that protect the inner workings of your cells — literally melt. Inside your cells, the proteins essential to life — the ones that extract energy from food or sunlight, fend off invaders, destroy waste products, and so on — often have beautifully precise shapes. The proteins start as long strands, then fold into helixes, hairpins, and other configurations, as dictated by the sequence of their components. These shapes define the function of proteins. But as the heat rises, the proteins unfold and the bonds that keep the structures together break: first the weaker ones, and then, as the temperature mounts, the stronger ones. At the most fundamental level, your body unravels.

At this point, no matter how strong or healthy you may be, your odds of survival are slim. The tiny tubes in your kidneys that filter out waste and impurities in your blood are collapsing. Muscle tissues are disintegrating. You develop holes in your intestines, and nasty toxins from your digestive tract flow into your bloodstream. Amid all this chaos, your circulatory system responds by clotting your blood, cutting off its flow to vital organs. This triggers what doctors call a clotting cascade, which uses up all the clotting proteins in your blood and, paradoxically, leaves you free to bleed elsewhere. Your insides melt and disintegrate — you are hemorrhaging everywhere.

After the bodies of the Gerrish-Chung family were airlifted off the mountain, Mariposa County Sheriff Jeremy Briese was faced with an obvious but perplexing question: What

— or who — killed them? Families don't just drop dead on hikes, especially families hiking with a young child. "There were no signs of trauma, no obvious cause of death. There was no suicide note," Mariposa County Sheriff's Office spokesperson Kristie Mitchell said. "They were out in the middle of a national forest on a day hike."

Media interest was intense. The Mariposa County Sheriff's Office parking lot was crowded with satellite trucks, and reporters were out hiking the trail, playing up the mystery of how the happy family could have died together on the trail.

"It could be a carbon monoxide situation. That's one of the reasons why we're treating it as a hazmat situation," Mitchell explained. According to one theory, the family could have been killed by a sudden release of poisonous gas from a nearby abandoned mine. A lightning strike had been ruled out — there was not a cloud in the sky that day, and there were no burn marks on the bodies.

Investigators also considered the possibility of toxic algae in the Merced River. Water samples from several locations on the river tested positive for Anatoxin A, a cynobacteria that can be deadly to animals but has never caused any recorded deaths among humans. And there was no evidence that

Gerrish, Chung, or their daughter had ingested any of the river water anyway. As for the idea of poisonous gas escaping from a mine, an old mine entrance was found two miles away, but law enforcement officials found no indication that the family had been anywhere near it.

Autopsy reports on Gerrish and Chung revealed little, as did a necropsy on Oski. Which is not surprising, because in most heat-related cases, people die of organ failure, which does not leave an easily detectable signature. Sometimes, an autopsy can find signs of internal bleeding or liver or kidney damage. In the case of Gerrish and Chung, determining the cause of death was complicated by the fact that their bodies had remained on the trail for some time after their deaths and were not well-preserved.

"I've never seen a death like this — they appeared to be a healthy family and a family canine," Briese told reporters. "Our hearts go out to the family, and we are working hard to provide closure. We're not going to rest until we figure this out."

Investigators hoped to find some clues in Gerrish's phone, but due to password protection, they couldn't get access to the contents immediately and had to send it to the FBI for help.

■ ■ ■ ■

Oski likely got into trouble first. Dogs are vulnerable to heat because they cannot sweat. As anyone who has ever taken a dog for a walk on a hot summer day knows, their only mechanism to release heat is panting, which is not particularly efficient. Some dogs are more vulnerable to heat than others. A recent study found that three main qualities corresponded with heat illness and death in dogs: weight, age, and body anatomy. Dogs with flat faces and wide skulls, such as English bulldogs, are twice as likely to succumb to heat as beagles, border collies, and other breeds with more pronounced snouts.

Not surprisingly, dogs with thick coats don't handle heat well. Golden retrievers are three times more likely than Labradors to suffer heat-related illness. Active, muscular dogs like greyhounds are also vulnerable. "Although Greyhounds have nice, long noses, thin hair, and are not usually associated with being overweight, they do have a high ratio of muscle," which, studies have shown, correlates to greater risk of heatstroke after exercising, one researcher said. "Also they are inclined to run about on the hottest days, without thought for the consequences."

Oski was a large, muscular dog with a thick coat. Climbing a steep slope on a 100-degree day would have been brutal.

One-year-old Miju would have felt the heat quickly too. She was in a carrier on her father's back, which was not a cool place to be. Besides being inside a fabric backpack, which would have trapped heat like an additional layer of clothing, she was inevitably absorbing heat from both her father's body and the sun. In addition, sweat glands in prepubescent children are not fully developed yet, which makes it difficult for them to release the heat that builds up in their bodies. Their bodies also contain a smaller volume of blood than adults', so when their heart pumps blood to their skin to try to cool down, it takes blood away from internal organs, potentially damaging them. This is why babies are so vulnerable when left alone in hot cars. They are basically defenseless against heat.

The heat would not have been easy for Gerrish either. He was a big guy, weighing about 210 pounds ("He had a dad-bod," his friend Steve Jeffe recalled fondly). Although all adult humans, no matter what sex or race, have about the same number of sweat glands, size is a disadvantage when trying to cool off, simply because a bigger animal carries a larger volume of heat in its body than a smaller animal. In addition, Gerrish was carrying Miju on his back, which added more weight, as well as inhibiting his ability to dissipate sweat.

Chung may have been in the best position to survive the climb. She was carrying only a light backpack with a hydration bladder and a few supplies for Miju. Given her smaller stature, she was likely slower to overheat. Although women have the same number of sweat glands as men, their variable hormone balance has a noticeable effect on sweat responses. For example, during the luteal phase of the menstrual cycle, which starts after ovulation and ends with the first day of a woman's period, women sweat similarly to men. But during the follicular stage, which starts on the last day of a woman's period and lasts until ovulation, women's sweat responses are initiated more slowly. Birth control pills can also increase core body temperature and make it more difficult for women to keep cool in the heat (or warm in the cold).

"Ellen was in great shape," one of her friends told me. "She exercised all the time. She didn't have an ounce of fat on her. If anyone could survive this, you would think it would be her."

■■■■

On October 21, a little more than two months after Gerrish, Chung, Miju, and their loyal pal Oski set out for a hike on a summer morning, Sheriff Briese held a press conference to reveal the official findings of the investigation into their deaths. "The cause of death was hyperthermia and probable dehydration due to environmental exposure," he said, voice trembling just a little. As in many heat-related deaths, there was no single piece of evidence that caused investigators to conclude that hyperthermia was the killer. It was based on scene investigation, the circumstances of their deaths, and the reasonable exclusion of other causes. Briese showed a graphic of the area where the bodies were found, noting that the trail's southern slope meant constant exposure to the sun and no shade. He estimated that the ground-level air temperature along the trail while the family was hiking was as high as 109 degrees.

What happened to Gerrish and Chung and Miju and Oski was not just a consequence of bad luck and poor decision-making in the wild. It was a tragedy that was shaped by our larger failure to reckon with the risks of life in a rapidly warming world, and with the nature of heat itself. We simply have not come to terms with it, especially in the way I am describing. It is not how anyone expects to die. In part, it's because we live in a technologically advanced world where it's all too easy to believe that the rough forces of nature have been tamed. But it's also because our world is changing so fast that we can't grasp the scale and urgency of the dangers we face.

In August 2022, more than a year after their deaths, family and friends gathered on a tranquil spot on one of Gerrish and Chung's properties in Mariposa to bury their ashes. It was a beautiful clear Saturday morning. But it was also "slightly surreal," Richard recalled. In the previous weeks, the area had been torched by the Oak Fire, which had burned 20,000 acres and 180 buildings and come within a half mile of Gerrish and Chung's home. There were still fire trucks parked on the side of the roads and bulldozers burying the last smoldering embers. It was the second time in four years this area had badly burned.

As family and friends stood nearby, Richard and Chung's sister Melissa placed a dark wooden box containing the commingled ashes of Gerrish, Chung, Miju, and Oski in the ground. Richard read from the journals of John Muir, who was born in Dunbar, Scotland, in 1838, not far from where Richard now lived. Muir's exuberant writing about the California mountains had led to the founding of America's national parks and inspired generations of people to think differently about their connection with nature. Muir was also a voice from a simpler time, when the biggest threat to his beloved Yosemite was a dam, not a fast-warming climate. In the passage Richard chose, Muir described death as a "home-going": "Myriads of rejoicing living creatures, daily, hourly, perhaps every moment sink into death's arms," Richard read. "Yet all enjoy life as we do, share Heaven's blessings with us, die and are buried in hallowed ground, come with us out of eternity and return into eternity."

Several other family members and friends read poems or offered remarks. Then Richard and Melissa covered the wooden box with a few shovelfuls of dirt and tamped it down. To the east, in the distance, the granite dome of El Capitan loomed. It wasn't noon yet, but already the heat was rising.

2. How Heat Shaped Us

To understand the dangers of extreme heat today, it helps to understand how we have lived with heat in the past. Among other things, we evolved clever ways to manage the heating and cooling of our bodies that gave our ancestors an evolutionary edge over competitors. To tell you about it, though, I have to go way back, because you can't separate heat from the beginning of things.

Fourteen billion years ago, the universe was compressed into a stupendously hot, incredibly dense nugget, which then rapidly expanded. This nugget cooled as it swelled; its particles gradually slowed their frenzied motion and aggregated into clumps, which over time formed stars, planets — and us.

How exactly life emerged out of the hot mess of the universe is only dimly understood. The most widely accepted theory is that life began around the volcanoes that rose above the ocean shortly after the earth formed, probably within the first hundred million years. The volcanoes were surrounded by hot geyser-fed ponds and bubbling hot springs, which were loaded with organic compounds from the asteroids and meteors that bombarded the planet. Volcanoes acted as chemical reactors, creating a hot volcanic soup. Somehow, RNA molecules grew, eventually growing longer and more complex and folding into true proteins and double-stranded DNA. They formed microbes that floated in thick mats on the volcanic ponds. When the ponds dried out, winds picked up their spores and spread them for miles. Rains eventually washed microbes into the ocean. "Once they reached the sea," science author Carl Zimmer writes, "the whole planet came alive."

Evolution's next trick was developing a way for animals to deal with temperature fluctuations. In the long arc of evolution, two strategies have emerged: one is to let your body's temperature change with the temperature around it, which is what creatures did for the first three and a half billion years or so. If necessary, these animals warmed themselves by basking in sunlight or sitting on warm rocks. This heat management strategy survives today in fish, frogs, lizards, alligators, and all the reptiles and amphibians.

Scientists call them ectotherms; you and I call them cold-blooded.

But around 260 million years ago, a new heat management strategy emerged. Some animals found a way to control their own internal temperature that was not dependent on the temperature of their environment. In effect, it turns their bodies into little heat engines, allowing them to operate independently of the world outside — as long as they can maintain a steady temperature inside. This heat management strategy remains alive and well in animals that scientists call endotherms but that you and I call warm-blooded: dogs, cats, whales, tigers, and virtually every other mammal on the planet, including us. Birds, which are flying dinosaurs, are also warm-blooded. ("Birds are not *like* flying dinosaurs," a scientist once corrected me. "They *are* flying dinosaurs.")

The birth of warm-bloodedness was an evolutionary leap, and one that scientists still don't fully understand. For one thing, the traits of warm-bloodedness do not transfer well to fossils, so you can't just look at the bones of a long-ago creature and determine whether it was warm- or cold-blooded. For another, the transition from cold-bloodedness to warm-bloodedness didn't happen with a single quick jump. Many species

— especially dinosaurs — had attributes of both.

At first glance, cold-blooded creatures seem to have it easy. Because they cannot regulate their body temperature internally, they spend thirty times less energy than warm-blooded creatures of the same size. So, while mammals and birds are constantly investing their calories in maintaining a high, stable body temperature, reptiles and amphibians can just search for a warm spot in their surrounding environment if they want to get cozy. But if cold-bloodedness is so great, why did mammals and birds develop a different strategy?

There are a lot of theories for why warm-blooded animals evolved high, stable body temperatures. To name a few: a stable body temperature aids physiological processes, such as digestion and the absorption of nutrients; it helps animals maintain activity over longer periods of time; it enables parents to take care of precocial offspring. Warm-bloodedness also allowed more precise and powerful functioning of certain cells in the nervous system, as well as in the heart and muscles.

Resistance to disease may have been another advantage. Insects bask in the sunlight to superheat their bodies and cook invading organisms; humans do the same by running a fever. But cold-blooded creatures depend on external sources of heat to kill invaders. If it's not hot out, a grasshopper can't fry the dangerous microbes in its body. And if that grasshopper goes looking for a spot of sunshine, it might venture into new places and get picked off by a predator. Warm-blooded animals don't have that risk. They can rev up the heat engine wherever they are.

Warm-blooded animals also move faster. John Grady, a biologist at the University of New Mexico, thinks the evolution to warm-bloodedness was accelerated by the competitive advantage that comes with being a speedy predator. Higher body temperature equals higher metabolism, which equals quicker reactions and more active predation. "Imagine an iguana the size of a cow," Grady told me. "These things existed. But they won't exist in today's world, because they are too slow. The closest thing we have are giant tortoises, and they have a strategy of just being armored. They don't have to be fast. When you are big, being fast is important. I think getting killed is a real problem if you are big and cold-blooded."

Whatever the particular advantages of warm-bloodedness may have been, it served mammals well. For the last seventy million years or so, they have spread across the globe, each creature a biological dynamo carrying its own fire inside. Their success gave rise, eventually, to a two-legged primate that developed a big brain and an even more sophisticated heat management system to go with it. To get a glimpse of this remarkable creature, just look in the mirror.

In 1974, a pile of bones was found in the Awash River valley in Ethiopia by Donald Johanson, who, at the time, was a professor at Case Western Reserve University in Ohio. The bones belonged to a female human ancestor who lived about 3.2 million years ago. Judging from her intact wisdom teeth and the shape of her hip bone, Johanson determined that she was a teenager when she died. He named her Lucy, after the Beatles' song "Lucy in the Sky with Diamonds," which Johanson and his team had been listening to in camp when she was found.

It was a remarkable discovery, rewriting the story of human evolution. Even at the time, Lucy was not the oldest human ancestor ever found, but she filled an important gap in the evolutionary tree from early hominins (that is, all our human ancestors since we diverged from apes about seven million years ago) to modern humans. She was also remarkably well preserved for a girl who had been buried for more than three million years. She had a spine, pelvis, and leg bones very similar to those of modern humans. She did not yet have the brain size of a modern human, but she was positively, indisputably bipedal.

It took a while for our ancestors to learn to stand up. From the structure and shape of the fossils they've left behind, paleontologists know that early hominins hung out mostly in trees. On the ground, they moved on all fours, not unlike the way chimps do today.

But Lucy was different. The shape of her lower femur, as well as the development of her knee, indicate that she walked upright at least part of the time. But she wasn't like us: she had wide hips and short legs. She was an evolutionary toddler just learning to venture out of the cover of the trees and onto the savanna.

The question is, what made Lucy stand up and start walking? It's a much-disputed subject among paleoanthropologists. Some argue that it allowed our ancestors to carry tools better. Others believe that it helped them reach fruit high in trees. Still others suggest that bipedalism was the basis for monogamy and family, in that it allowed male hominins to go out and get food, which the female chimps would reward with companionship and sex.

Or standing up may have been a way of keeping cool. It allowed Lucy to catch breezes and help body heat dissipate more easily. It also got her up off the ground, which is always significantly warmer than the air a few feet above it.

Whatever her motivation may have been, Lucy walked. And it changed everything.

To understand the power of heat, you have to think of it not just as a change in temperature, but as an evolutionary hurdle. Heat management is a survival tool for all life on Earth, and the strategies to deal with it are as diverse and colorful as the animal kingdom itself.

Elephants are particularly fascinating. They spend a lot of time in the sun. To cool off, they seek shade and water. (In Botswana, I once watched a young elephant frolic in a muddy watering hole on a hot day like a six-year-old kid at summer camp.) Their thin hair and flapping ears help with heat dissipation. More important, as temperatures rise, their hides become more permeable. Their skin effectively opens up, allowing them to perspire, even though they don't actually have sweat glands.

Koalas hug trees with bark that is cooler than the air temperature. Kangaroos spit on their arms to wet them and cool off. Some squirrels use their bushy tails as parasols. Hippos roll in mud (water evaporates more slowly from mud, keeping them cool longer). Lions climb trees to get off the hot ground. Rabbits send blood to their big ears, using them as radiators. Vultures and storks defecate on their legs. Herons, nighthawks, pelicans, doves, and owls cool themselves with gular fluttering, a frequent vibration of their throat membranes, which increases airflow and thus increases evaporation. Giraffes' beautifully patterned skin functions like a network of thermal windows. They direct warm blood to the vessels at the edges of the spots, forcing heat out of the animals' bodies.

Other animals build structures to cool themselves, in some ways not so different from the way humans construct air-conditioned buildings. Termites build an elaborate system of air pockets within their mounds. Bees harvest water when they're on their travels, then return to the hive and pass it by mouth to hive bees, which spread the droplets on the honeycomb. Other bees fan the water with their wings to cool the hive.

The silver Saharan ant, which thrives in

a scorching environment in the Sahara and on the Arabian Peninsula, has evolved some remarkable heat-coping strategies. When a silver Saharan ant goes in search of food, it has ten minutes before the desert heat will literally fry it. It ventures out in temperatures above 122 degrees, usually to scavenge the corpses of heat-killed animals. The ants go out only when it's too hot for the lizards that prey on them but still cool enough that they don't get cooked instantly. To avoid the superhot sand, they run fast — up to a meter per second, which, given their small body size, is the equivalent of a human running 450 miles per hour. The beautiful silver hue of the ant comes from unique triangular hairs on its body, which reflect away heat (just as wearing a white shirt keeps you cooler than wearing a black shirt).

But you can't talk about heat and animals without talking about camels. Not long ago, I spent a few days riding an Arabian camel across the Wadi Rum, a desert region in Jordan. To anyone who is used to riding a horse, camels are strange, calm, smelly, uncharismatic animals. The one I rode didn't really seem to care much whether I was on him or not. My Jordanian guide said the camel did not have a name, and the guide did not seem to have much affection for him, which just made me feel sorry for the poor beast having to trudge across the desert with me on his back.

Camels evolved in North America roughly forty million years ago, and their best-known features — their long eyelashes, their wide feet, their humps — may have come in response to North American winters. They crossed the land bridge at the Bering Strait about fourteen thousand years ago, ending up on the Arabian Peninsula, among other places. They have been domesticated for thousands of years, nearly as long as horses.

Despite (or perhaps because of) their origins, camels are extremely well adapted to life in hot deserts. They have translucent eyelids that allow them to close their eyes and keep walking during sandstorms. They can also close their nose to keep sand out and water in. Thick tissue over a camel's sternum allows it to easily raise its head above the hot ground when it is lying down. A camel's hump shades and insulates internal organs from heat. Contrary to some myths, the hump doesn't store water, it stores fat — which it can tap into when food is scarce. When a camel doesn't eat for a long time, its hump sags.

For all living things, life in a hot climate requires careful water management, and camels are supremely good at it. They have unusual oval blood cells, which can circulate through thick blood, and quickly expand when water is available. During the winter and cold seasons, a camel can go without water for months. Under very hot conditions, it may drink only every eight to ten days and lose up to a third of its body weight through dehydration. A dehydrated camel urinates only drops of concentrated urine, which look like white stripes on its hind legs and tail (it's actually salt crystals). The way camels process their urine not only conserves water, but it also allows them to drink water that is even saltier than seawater, and to eat salty plants that would be toxic to most other animals. Their feces are so dry, they can fuel fires.

There aren't many people who have thought more about heat as an evolutionary force than Jill Pruetz. For the past twenty years, she has spent a good part of every year in Senegal, near the village of Fongoli, where she has been studying chimpanzees that live in a hot environment. Pruetz has a way of talking about being among the chimps that suggests she knows them better than many people know their own children.

Pruetz and I met on a sunny spring day at

a restaurant in Bastrop, Texas, near where she lives on a five-acre farm. She grew up in south Texas and became fascinated with chimpanzees shortly after college, when she went to work at a chimpanzee center that bred chimps for biomedical research. She is now a professor of anthropology at Texas State, and runs the Fongoli Savanna Chimpanzee Project, where thirty-two chimps live in a 100-square-kilometer area, outside the national park.

Pruetz and I sat at a wooden picnic table above the Colorado River and ate pizza while we talked. "I study chimps for a lot of reasons," she told me. "But mostly because they are our closest living relative, and we can learn a lot about early human development by looking at how chimps behave and react to different kinds of stress in their lives."

For the Fongoli chimps, heat is extremely stressful. During the hot dry season in Senegal, which peaks in March and April, the temperature can hit 120 degrees. "The heat is like a slap in the face," Pruetz said. Trees are leafless. Water is scarce. Fires burn across their territory. These chimps live in the hottest, most arid place that chimps are known to exist. It is a brutal, apocalyptic landscape that is totally unlike the lush forests and jungles that every other chimp on the planet inhabits.

The chimps have been living on this piece of turf for a very long time. "Millennia," Pruetz told me. Over time, the chimps gradually evolved a catalogue of strange behaviors — ones rarely if ever seen in others. Forest chimpanzees get enough water from the fruit in their diet, so they need less drinking water and can wander in search of food. The Fongoli chimps, by contrast, require daily drinking water and anchor themselves to reliable water sources in the arid landscape.

And while forest chimpanzees are active throughout the day, Pruetz found that the savanna chimpanzees rest for five to seven hours. Pruetz could often find them lurking in small caves in the dry season, and when the rainy season arrived, the chimpanzees would slip into newly formed ponds and bob there for hours. Forest chimpanzees typically spend all night in nests they build in trees. But at Fongoli, the research team noticed that the chimpanzees often made a late-night racket.

"During the hot season, the chimps totally change their behavior," Pruetz told me. They stare at the sky, waiting for the rain they know is coming. At Fongoli, there are few trees, and the ones that are there don't have many leaves for shade. On one particularly hot day a few years ago, Pruetz watched an adolescent chimp hiding in the shadow of a single tree trunk. As the day passed, the chimp moved with the shadow, trying to escape the heat.

Pruetz has also noticed something else, something that was perhaps key to the whole human story: in the heat, the Fongoli chimps spend more time standing up and walking around than chimps that live in cooler places.

Lucy lived in a rapidly changing world. It was nowhere near as rapidly changing as ours is today, but in evolutionary terms, it was on the move. The climate of East Africa was growing hotter and drier. Rain forests gave way to woodlands, and as the landscape opened up, the savanna emerged. "Over the past three to four million years, the scenery of East Africa shifted from the set of Tarzan to that of *The Lion King,*" writes Lewis Dartnell in *Origins: How Earth's History Shaped Human History.* Ethiopia's Rift Valley became a very complex environment, with woods and highlands, ridges, steep escarpments, hills, plateaus and plains, valleys, and deep freshwater lakes on the floor of the rift, which was gradually widening.

Meanwhile, volcanoes like Mount Kilimanjaro were spewing pumice and ash across the whole region. New species like zebra were emerging from under the trees and appearing in the grasslands.

In this dynamic new world, Lucy had to be nimble. Water supplies were drying up and filling again with each passing rainstorm. Leopards and lions lurked in the ravines — she was both predator and prey (we think of the world that she lived in as so different from ours, but in fact, the creatures that made up East Africa at that time were similar to what is there today — lions and hyenas and elephants were all more or less the same). If the behavior of chimps today is any indication, these early hominins were wary of open ground, fleeing back into the safety of trees whenever they could. The changing terrain, and the need to navigate through it, meant that the most vulnerable among them were killed by predators. But the most adaptive ones survived and learned new skills, including hunting with tools, which helped them shift away from a diet of fruit and termites and small forest creatures to a more meat-centric diet, including gazelle and zebra, which they might have hunted in groups.

Kevin Hunt, a professor of anthropology

at Indiana University who studies human evolution, believes bipedalism likely evolved gradually, over a million years or so. Lucy was an example of the first phase — she may have stood up both to escape the heat and to help her reach for fruit. The second phase, marked by the arrival of *Homo erectus,* had elongated limbs that allowed them to walk and run faster, a more slender body that better dissipated heat, and a more carnivorous diet.

But to take the next step in human evolution, to really allow our ancestors to range widely in the newly warmed world, they still needed one more key evolutionary innovation. They needed to learn how to sweat.

In our human ancestors, the evolution of the sweat gland is even more complex than the evolution of bipedalism. Bipedalism can be deduced from fossil bones. Sweat glands can't. What is known about them can only be inferred by hints of behavior patterns found in other ways, and by the evidence we see in our own bodies and in the bodies of other animals.

What is clear is that as Lucy and her generation made their way out of the trees and into the savanna, they had to contend with heat in a way that they never did when living in the trees. In both cases, our ancestors came up with important innovations that still have big implications for how we live today.

First, there was sunlight to deal with. As they wandered out from under the trees, our ancestors were exposed to more and more ultraviolet radiation, which damages the cellular structure of skin and can harm DNA. So Lucy and her ancestors evolved the ability to produce melanin, the dark-brown pigment that acts as a natural sunscreen. For several million years, our ancestors were all dark-skinned. It was only after they migrated out of Africa and settled in more northern climates, and in high latitudes, that dark skin became an evolutionary disadvantage because it limited the sunlight getting through to trigger the production of vitamin D. So in regions where the sunlight was less intense, lighter skin had an advantage.

Dealing with heat was more complex. In warm-blooded animals, more sunlight means more heat. More activity means more heat too. How far you can chase a wounded antelope in the heat depends on how well you manage heat. On the African savanna if you overheat, you go hungry. In addition, our ancestors' brains were evolving, and getting bigger. But big brains require a lot of cooling, and so developing a robust cooling system was important to advancing other skills, such as toolmaking.

The solution that evolution came up with was to build what amounts to an internal sprinkler system that douses our skin with water when we get too hot. As the water evaporates, it carries the heat off with it, cooling off our skin and the blood circulating just below it. When that cooler blood circulates, it brings down the temperature in our bodies.

If you've ever ridden a horse on a hot day, you know that other animals sweat. Horses, as well as many other mammals, have a particular kind of sweat gland that is part of their hair follicles called an apocrine gland. It sends out a thick, milky white liquid. You see it most clearly on racehorses, which sometimes finish a race looking like their necks are covered in shaving cream (thus the origin of the phrase "get in a lather"). Many furred mammals have apocrine glands, including camels and donkeys, as well as chimpanzees. These glands help with heat management, but they can't really dissipate a lot of heat fast.

Humans have some apocrine glands in our armpits and pubic areas, which are evolutionary leftovers from an earlier time.

They respond to nerves as well as heat, and are why your armpits sweat in an interview, and also why your sweat has a particular odor. Some anthropologists think that smell is an ancient sexual attractant; that it's one of the ways we got to know one another.

But while our ancestors were hunting for food in the heat on the African savanna, they also perfected a much better heat management tool, which is the eccrine sweat gland. Instead of creating a lather, it is basically a mechanism to squirt water on your body, which will then evaporate and cool you off. It's simple but brilliant. Hominins didn't invent the eccrine gland. Old World monkeys like macaques have equal parts eccrine and apocrine glands. Our closer relatives, chimpanzees and gorillas, bear roughly two eccrine for every one apocrine gland. But beyond the apocrine leftovers in our armpits and pubic areas, human sweat glands are all eccrine.

Today, you and I have about two million of these sweat glands on our body. The glands themselves are like little coiled tubes buried in your skin. They are tiny, the size of a cell — you need a microscope to see them. They are not evenly distributed on your body: you have the most sweat glands on your hands, feet, and face, and the least on your butt. Sex differences are small. Women often have more sweat glands in any given area than men, but men often have a higher maximum sweating rate. The liquid the glands secrete is 99.5 percent water — its only function is to wet your skin. In hot weather, most people can easily sweat one quart per hour or 12 quarts a day, which is about ten times more than a chimp sweats.

To make our sweat glands even more effective, however, Lucy's offspring made another evolutionary adjustment: they lost their body hair. For the evaporative sweat to really work, hair (or fur, which is just another name for hair on nonhuman animals) gets in the way, matting down when wet and interfering with the efficient transfer of heat away from your body. The only place we still have significant hair is on our heads, and that's because our brains are so sensitive to heat, and in this situation, hair works as a sunshade to help keep our brains cool. (It also adds cushioning in a fall.)

The loss of hair on our bodies and the development of eccrine sweat glands were important evolutionary events, perhaps as important as the use of tools or fire. Other animals on the African savanna had developed heat stress strategies — the simplest of which is panting. But for a predator, panting is not a great strategy. A lion or a hyena can move very fast for short distances, but it can't pant while it runs. In the heat, it has to stop, rest, pant, and recover its thermal equilibrium. Humans figured out a way to keep cool in motion. We don't have to stop and pant. We sweat as we go. In the story of human evolution, this was a very big deal. By managing heat, humans were able to go farther from water holes, begin long-distance travel, and expand their hunting range.

Humans became excellent hot-weather hunters. They could venture out in the heat of the day when other animals couldn't, giving them a predatory advantage. By the time *Homo erectus* appeared about two million years ago, our ancestors were on their way to becoming endurance athletes, with long legs, nimble feet, and strong leg and hip muscles. With their superior heat management systems, they could literally run down an animal until it has heatstroke. This practice continues today. In the Kalahari Desert in southern Africa, modern hunter-gatherers are able to kill a kudu, a kind of antelope that is far faster than humans over short distances, by chasing it for hours in the middle of the hot day, until it literally collapses of heat exhaustion.*

But the heat management strategies of humans, like those of all living things, have been optimized for the Goldilocks Zone we have been living in for the last ten thousand years or so. They are a holdover from a world that is changing fast — far too fast for evolutionary selection to keep up. When it comes to heat management, we are like actors in Hollywood's silent era who suddenly find themselves cast in speaking roles. We know the script, but our skills are no longer well-matched for the world we live in.

* There is a common but false belief that our ancestors in hot climates developed a taste for spicy foods because it made them sweat. In fact, a taste for spicy foods likely arose because before refrigeration, spices worked as a food preservative, which is particularly important in hot places where food spoils quickly. Garlic, onion, allspice and oregano are excellent bacteria killers, as are thyme, cinnamon, tarragon and cumin. "People who enjoyed food with antibacterial spices probably were healthier, especially in hot climates," evolutionary biologist Paul Sherman said. "They lived longer and left more offspring. And they taught their offspring: 'This is how to cook a mastodon.' "

3. Heat Islands

On a scorching day in downtown Phoenix, when the temperature soars to 115 degrees or more, the sunshine assaults you, pushing you to seek cover. The air feels solid, a hazy, ozone-soaked curtain of heat. You feel it radiating up from the parking lot through your shoes. Metal bus stops on Van Buren Street become convection ovens. Flights are delayed at Sky Harbor International Airport because the planes can't get enough lift in the thin, hot air. At City Hall, where the entrance to the building is emblazoned with a giant metallic emblem of the sun, workers eat lunch in the lobby rather than venture outside to nearby restaurants. On the outskirts of the city, power lines sag and buzz, overloaded with electrons as the demand for air-conditioning soars and the entire grid is pushed to the max. In an Arizona heat wave, electricity is not a convenience. It is a tool for survival.

Modern cities are empires of asphalt and

concrete and steel, materials that absorb and amplify heat during the day, then radiate it out at night. Air conditioners exhaust hot air, exacerbating the problem of urban heat buildup. Downtown Phoenix can be as much as twenty degrees hotter than the surrounding area. New York City averages two to five degrees warmer than its leafy suburbs during the day — and sometimes twenty degrees warmer at night. This phenomenon, known to city planners and heat researchers as the urban heat island effect, is so pervasive that climate skeptics once claimed that climate change is merely an illusion created by thousands of once-rural meteorological stations becoming surrounded by urban development (the argument, like most arguments pushed by climate skeptics, has been thoroughly debunked).

"For cities, the urban heat island effect has had far more impact on local temperatures than climate change itself," said David Hondula, a scientist at Arizona State University and the director of heat response and mitigation for the city of Phoenix.

In 2021, there were 339 heat-related deaths in Maricopa County. That might not sound like a lot, especially compared to the 70,000 people who died in just a few days during a heat wave in Europe in 2003, but it was more than triple the number from a decade earlier, before the city had begun taking deliberate action against heat. And the trend fit with the growing number of heat-related deaths in cities. A recent study by the National Academy of Sciences found that globally, the heat risk in urban areas tripled over the last forty years, putting 1.7 billion people at risk. Unless we take dramatic action to reduce CO_2 pollution and change how we live, the number of people at risk will grow exponentially. By 2050, 70 percent of the world's population will live in cities.

The cascading dangers of extreme heat in urban areas are just beginning to be understood, even in places like Phoenix, which has long been one of the hottest big cities in America. To Mikhail Chester, the director of the Metis Center for Infrastructure and Sustainable Engineering at Arizona State University (ASU), the risk of a heat-driven catastrophe increases every year. "What will the Hurricane Katrina of extreme heat look like?" he wondered aloud as we sat in a café near the ASU campus a few years ago. Hurricane Katrina, which hit New Orleans in 2005, resulting in nearly two thousand deaths and more than $100 billion in economic damage, demonstrated just how unprepared a city can be for extreme climate events.

"Hurricane Katrina caused a cascading failure of urban infrastructure in New Orleans that no one really predicted," Chester explained. "Levees broke. People were stranded. Rescue operations failed. Extreme heat could lead to a similar cascading failure in Phoenix, exposing vulnerabilities and weaknesses in the region's infrastructure that are difficult to foresee."

In Chester's view, a Phoenix heat catastrophe begins with a blackout. It could be triggered any number of ways. On a hot day, a wildfire knocks out a major power line. A substation blows. A Russian hacker crashes the grid. In 2011, a utility worker doing routine maintenance near Yuma caused the loss of a 500-kilavolt power line that began a rolling blackout that shut off power for twelve hours to seven million people, including virtually the entire city of San Diego, causing economic losses of $100 million. "A major blackout in Phoenix could easily cost billions," said Chester.

When a city like Phoenix goes dark, the comforts and conveniences of modern life fray. Without air-conditioning, temperatures in homes and office buildings soar. (Ironically, new, highly efficient LEED-certified buildings are tightly sealed, making them dangerous heat traps when the power goes out.) Traffic signals fail. Highways gridlock with people fleeing the overheated city. Without power, gas pumps don't work, leaving vehicles stranded with empty tanks. Underground water pipes crack from the heat, and water pumps fail, leaving people scrounging for fresh water. Hospitals overflow with people suffering from heat exhaustion and heatstroke. If there are wildfires in the nearby mountains, the air will become hazy and difficult to breathe. If the blackout continues for more than thirty-six hours or so, it's likely the National Guard will need to be called in to maintain order and control widespread looting and general mayhem.

And people will start dying. How many? "Katrina-like numbers," Chester predicted. Which is to say, thousands.

In the café, Chester described all this coolly, as if a Phoenix heat apocalypse is a matter of fact, not hypothesis.

"How likely is this to happen?" I asked.

"I'd say the odds are about the same as the odds that another big hurricane will hit New Orleans," Chester explained. "It's more a question of when, not if."

A hot city is different than a hot jungle or a hot desert. Urban heat feels crueler and more intimate than the heat you feel in nature. Despite the fact that cities are full of people, urban heat has the perverse effect of creating islands of isolation and hardship for anyone without the means or the social connections to access cool spaces. It makes the hardship of poverty harder and turns even the simplest tasks of daily life into risky adventures.

Consider Anjalai (she asked that her last name not be used), who is thirty-nine years old, with broad shoulders and a kind of eagerness in her eyes that suggests she's always thinking of the next question to ask. She has a gold ring in her right nostril, and often wears gold earrings with a small pearl dangling at the end of a tiny gold chain. She lives with her seventeen-year-old daughter and forty-nine-year-old husband in a hut with a palm-leaf roof in the Ramapuram neighborhood of Chennai, a city of eleven million in southern India. Chennai is home to some of the richest entrepreneurs and businesspeople in India; it is also home to more than a million people like Anjalai who live in slums that are wildernesses of discarded plastic and hungry dogs.

Anjalai's hut is small, about three hundred square feet. It is neat and tidy, with a single circulating fan on the ceiling. Chennai, which is not far from the equator, is hot and humid most of the year. But May is a particularly brutal month. The temperature during the day is almost always 90 degrees or above, and it barely cools off at night. There's none of the sharpness of the dry desert heat in Phoenix. In Chennai, your sweat doesn't evaporate, it just pools around you. It's jungle heat, thick and heavy, although the jungle that once covered this landscape is just a memory, long paved over and concreted. The big tamarind trees, the coconut palms, the banana trees, the neem trees with their long, slender leaves — they're mostly all gone. The percentage of tree cover in Chennai is now about the same as in Phoenix.

During May, when I met her, Anjalai's daily routines were defined and driven by heat. In the morning, she ate cold rice porridge, which she believes helps cool her body. Every day, before she left for work, she wetted down the thatched roof of their hut, then spread water around the dirt at the base of the hut. The dampness, she said, helps absorb the heat. She moved slowly, preserving her strength for the day.

"How are you feeling, Anan?" she always asked her husband.

Anjalai worried about him every day. He has a heart condition, and the stress of the heat is hard on him. He works in construction, which means he is outdoors most of the day in the sun. He gets no respite in air-conditioning, no cool water to dunk himself in. She would like him to stay home, but they need the money, so he usually works a few days a week to help pay the bills. They cannot afford a mobile phone, so Anjalai fears that if something happens to him, she won't hear about it for many hours. "Today, he stayed home," Anjalai told me one day, and I could hear the relief in her voice.

At around 11 a.m. every day except Sunday, she goes to work cleaning houses. She rides through the city on her rusted-out one-speed bike with tires whitened by the sun. She has five or six houses she cleans every week, in a random rotation. They are rich people's houses, by which Anjalai means they are houses of people who have jobs and, therefore, houses with nice windows and big rooms and air-conditioning. Working indoors gives her some relief. She often thinks of her husband, and feels guilty that she can cool off but he can't. But that feeling doesn't last too long. Inevitably, she must go up on the roof to clean up there too (many rooftops in Chennai double as living spaces). The sun burns in the sky and the air is thick and heavy. "It almost hurts to be up there," she told me. Sometimes when she is working on the roof, she thinks about the village where she grew up, in a rural area outside of Chennai, where there were trees that she could sit beneath for shade, watching the branches blow in the wind and eating pieces of cool coconut. By the time she was twenty, she and her husband were married and her family moved to the city to seek jobs and try to build a better life.

When they arrived in 2004, Chennai was booming, with thriving auto, health care, tech, and film industries. There were still a few signs of old Madras, as the city used to be known, like the Ice House, built in 1842 by American businessman Frederic Tudor. Tudor cut blocks of ice out of lakes and ponds in New England, packed them in sawdust, then shipped them around the world. New England ice was wildly popular among the Brits living in Madras, who loved to sip gin and tonics under the banana trees. Tudor's business collapsed with the coming of ice machines and other modern conveniences, but the Ice House still stands in Chennai, down near the beach.

Back in those days, Chennai was a gentler city. The roads were dirt, still shaded by patches of jungle. Homes and buildings had thick roofs made of brick and wood and plaster, known as Madras roofs, that helped to keep them cool. Streets were arranged to capture the breeze that comes in off the Bay of Bengal. Buildings were deliberately constructed with space between them, so air could flow freely. There was plenty of water, most of it drawn from neighborhood wells. It was hot, but you drank buttermilk or, if you were lucky, gin and tonics. You moved slowly and you coped with it. It was life in the tropics.

But then the great urbanization of India began in the 1970s. Unlike Delhi and other cities, which went vertical, Chennai spread horizontally. Wetlands and swamps were covered over. The arrival of air-conditioning meant that when they built things now, developers and city officials didn't bother to think about sea breezes or airflow. Traditional wells were replaced with deep-bore wells, which were dependent on the region's deep aquifer, which, among other things, was prone to saltwater intrusions that made it undrinkable.

Chennai is now India's sixth-largest city, with a population five times larger than Paris. Nearly a hundred square miles of land has been paved over or developed. Eighty percent of the wetlands are gone.

The costs of this development have become apparent — not just with heat, but with water. In 2015, after days of violent rainfall, the concrete and asphalt channeled all the water into the city and nearly drowned it. Then in 2019, much of India baked under a heat wave with temperatures as high as 123 degrees. In Chennai, the heat was made all the more brutal by the fact that it was running out of water. The city gets an average of about fifty-five inches of rainfall a year, more than twice the amount that falls on London. Yet in 2019, due to poor water storage, people in the city did not have enough to drink. Ten million liters of water were trucked in each day until the heat faded. As one journalist wrote, "The ancient south Indian port has become a case study in what can go wrong when industrialization, urbanization and extreme weather converge and a booming metropolis paves over its flood plain to satisfy demand for new homes, factories and offices."

After Anjalai finished cleaning the rooftop, she rode her bike a few miles through the crowded streets to the Pudiyador school, a private school started twenty years ago by a university professor to help slum kids prep for college. Anjalai had started working there years ago, helping to clean the buildings.

But the school administrators noticed she was good with kids and loved to learn, so they gave her a job as a part-time teacher. And that worked out so well that she became a full-time teacher, working from 4 to 8 p.m. every day with a group of seven- and eight-year-olds. For this, she earns a paycheck of $70 per month.

As she pedaled through the city one day in late May, her gold sari flowing out behind her, the afternoon heat parted like water in front of her. May brings *Agni Nakshatram* (scorching star), a celebration of early summer and of Lord Murugan, a Hindu god and god of war. The intense heat of May is sometimes called *Kathiri Veyil* (scissor veil), because the sun's rays feel as sharp as scissors on your skin. Traditionally, people in Chennai avoid housewarmings, marriages, and other gatherings during this time. They stop eating meat and drinking refrigerated water. Instead, they drink sweet warm water and lemon juice, or water suffused with cumin seeds. One physician I talked to recommended oil baths twice a week, and drinking fenugreek water (to make fenugreek water, fifteen fenugreek seeds, which smell like maple syrup and have long been used in India as a medicinal herb, are dropped into a glass of water at night, then consumed the next morning).*

When Anjalai arrived at Pudiyador on her bike, she was sweaty, her face flushed, but she did not complain. Because of the Covid-19 pandemic, the classroom was empty — as for many other teachers in the world, teaching had gone virtual. She grabbed a laptop out of a locked closet, then sat on a mat on the floor in the center of the pink room, her legs crossed. A fan spun above her. The school administrators refused to install air-conditioning because they fear it will spoil kids and make it more difficult for them to cope with the heat when they are at home.

Anjalai spent the next four hours sitting on the floor staring at her computer, talking with kids in Tamil about their math and geography assignments. The lights flickered. The connection on her laptop failed several times, and she had to reconnect.

A little after 8 p.m., she finished. She closed the laptop and stowed it again in the locked closet. Then she climbed back on her bike and pedaled home through the steamy night. Dogs barked. Men squatted in a circle in the street, talking quietly. The air

* All of these remedies, while culturally important, have little or no scientific basis.

smelled of rot, of drying laundry, of jasmine, of stagnant water. There would be no break from the heat in the darkness, just the heavy weight of it all night long, until the sun came up and the day began again.

Cities like Phoenix have what are known as cooling centers that open when the heat is on. Often they are libraries or community centers with air-conditioning and snacks and cold water to drink. In theory, they offer refuge from the heat for people who have no refuge of their own. Most of the cooling centers I've visited have only had a handful of people in them, which, when you think about it, is not surprising. The people who need them most are often the people who don't have the means to get there.

So some Good Samaritans take to the streets to meet people where they are. One hot afternoon, I drove around in Glendale, a town just outside of Phoenix, with Brian Faretta and Rich Heitz, who were both ministers in training at the Phoenix Rescue Mission, a charitable church-based group dedicated to getting people off the streets. The group had just launched an initiative called Code: Red to pass out water and other essentials to people on the street during heat waves. "Our strategy is simple," Heitz explained to me. "We find people and give them water and try to get them out of the heat."

Heitz, forty-eight, has lived in Arizona most of his adult life. He is a gentle man with a goatee and a Harley-Davidson cap (when he showed me a picture of his '99 Harley Road King, he said, only half joking, "This is God"). Before he joined Phoenix Rescue Mission, Heitz spent seven years on the streets of Phoenix as a heroin addict. "I lost myself in numbness," he says. He spent a few years in jail for various charges and has now gone clean and is devoting his life to helping others do the same.

We pulled into Sands Park, a typical suburban rec area with basketball courts and picnic areas. Heitz and Faretta headed over to a concrete bathroom, where they found a middle-aged woman sitting in the shade on the floor near the entrance.

She had sunburned brown skin and long gray hair and a pleasant smile. She was dressed in soiled jeans and a T-shirt. Beside her was what looked like a children's coloring book. On the cover, written in red crayon, were the words "It's Raining Love."

"How are you doing, Sherri?" Heitz asked her. "You doing okay in the heat?"

I noticed her face was flushed, and there were rings of sweat under her arms.

"Yeah, I'm keeping cool."

Heitz offered her a couple of bottles of water, which she took, stockpiling them beside her.

As we walked back to the van, Heitz said this summer would be brutal for her and for all the homeless in the city. "If you're smart, you figure out ways to survive, to adapt," he said. "You find friends with cool houses where you can crash during the day. You learn which churches are open."

But not everyone figures it out. Heitz told me about a man he found lying in the heat on the sidewalk last summer. His face was flushed, his eyes were dilated, and he wasn't moving. "I called 911, and they took him to the hospital," Heitz said. "The guy was cooking right there on the sidewalk."

Climate change compounds risks for cities: heat, floods, failing infrastructure, displaced people. After the floods in 2015, which submerged large parts of Chennai, the Tamil Nadu Urban Habitat Development Board in Chennai began moving people out of huts and shacks in low-lying areas of the city and forcibly moving them to "safer" housing. One of these people was named Mercy

Muthu. She and her family were relocated to a high-rise building about ten miles away, in a development called Perumbakkam.

Perumbakkam is a world unto itself, a collection of concrete high-rises in a forgotten corner of the city. Imagine the most derelict low-income federal housing project you have ever seen, and then imagine something worse. They are concrete towers ten stories tall, with nothing but weeds and broken concrete walks between them. Not only are there no trees, there are also no bushes or shrubs. In the summer, when I talked with Muthu and her family, it was just a relentless landscape of heat.

Muthu has three kids, all teenagers. She is forty-one, a strong-looking woman dressed in a blue and gold sari, her hair pulled back in a tight bun. Her apartment is on the fifth floor of building 3, in the center of Perumbakkam. Inside, the walls are painted a cheerful pink. It is crowded — one bedroom, a kitchen, a living room. The only window is in the bedroom, where they all sleep on the floor. The kids play in the concrete corridors. "It is too hot for them to go outside," Muthu told me. "They all get heat rashes. At least where we lived before, they could go out and play."

The isolation of Perumbakkam, so far

from the center of the city, exacerbates the heat risk. It is a heat island unto itself. When Muthu and her family walk to the shops, which are nearly a mile away, they have to take plenty of water with them, as if they were setting out across a desert. Some people faint. Muthu's husband works as a driver for a taxi company. In the neighborhood where they used to live, they were a five-minute walk from the auto stand. "Now, he has to travel thirty-two kilometers one way, sixty kilometers round trip, just to go and try to make money," she told me. The isolation makes it harder for women to get jobs too. Many women find employment as domestic workers, and where they used to live, they could walk to the houses where they work. Now, they can't. Muthu told me about a friend of hers who gets up at 2 a.m. every day, packs lunch by 3 a.m., boards the bus by 4 a.m., works until 1 p.m., then arrives back home at 3 p.m. "The stress, combined with the heat, has given her boils all over her body," Muthu told me.

Many of the apartments do not have running water. Muthu is lucky — hers does, but it smells bad and anyone who drinks it gets sick, so they only use it for washing (In India, like many poor countries, clean drinking water is not a widely accessible public good.)

For drinking, Muthu and her kids go down to the well in the courtyard and carry the water back in orange plastic jugs. She constantly has to warn her kids not to touch any of the electrical equipment around the water pumps. A few weeks before I met her, a child had been electrocuted by touching a wire.

Muthu has no air-conditioning. Of the six hundred or so apartments in Perumbakkam, she estimated that three or four have air-conditioning. Even if Muthu had it, she said, they couldn't afford to run it.

"I feel bad for the old people," Muthu told me. "They are afraid to go out."

The elevator is a particularly spooky risk. "When the power goes out, it stops," she said. "The old people can be stuck for many hours. It happens all the time. The alarm sounds, but it takes a while to find out where they are, and then how do you get them out? It's a very difficult situation. No one wants to get cooked in an elevator."

In Phoenix, temperature is a signifier of class, wealth, and, often, race. If you're rich, you have a big house with enough air-conditioning to chill a martini. And if you are poor, like Leonor Juarez, a forty-six-year-old single mother of four whom I met on a July afternoon when the temperature was hovering around 115 degrees, you live in South Phoenix where trees are few and you hope you can squeeze enough money out of your weekly paycheck to run the air-conditioning for a few hours on summer nights.

On hot days, Juarez's small apartment feels like a cave, not so different from Muthu's concrete apartment in Chennai. She has heavy purple curtains on the windows to block the sun. "I could not live here without air-conditioning," she told me. Because she is poor and has no credit, she doesn't qualify for the usual monthly billing from Salt River Project (SRP), her electric utility. Instead, to pay for electricity and keep her air-conditioning running, SRP gave her a card reader that plugs into an outlet that she has to feed like a jukebox to keep the power on. Juarez only turns on her air conditioner a few hours a day — still, her electric bill can run $500 a month during the summer, which is more than she pays for rent. To Juarez, who takes a bus five miles to a laundromat in the middle of the night because washing machines are discounted to 50 cents a load after 1 a.m., $500 is a tremendous amount of money.

Juarez showed me the meter on the card reader: she had $49 worth of credit on it, enough for a few more days of power. And when that runs out? "I am in trouble," she said bluntly. Juarez, who works several days a week as an in-home caretaker for the elderly, said she knows of several people who lived alone and died because they failed to pay their electric bills and tried to live without air-conditioning.

One such woman was named Stephanie Pullman. She was seventy-two years old and lived alone in a small house in Sun City West, a development north of downtown Phoenix. After raising four kids in Ohio, Pullman had moved to Arizona in 1988 to escape the cold Ohio winters. She worked in a hospital, then retired in 2011 and lived on a fixed income of less than $1,000 a month. During the summer of 2018, she was late to pay her electric bill and owed $176.84. On September fifth, Pullman paid $125, leaving $51.84 unpaid. Two days later, when the temperature hit 107 degrees, her electric company, Arizona Public Service (APS), cut off her power. A week later, Pullman's daughter became worried when she hadn't heard from her mother and alerted locals. A Maricopa County sheriff entered the house and found Pullman dead in her bed. Cause of death: heat exposure.

A few months later, *Phoenix New Times* published an account of Pullman's death, including the fact that APS had turned off Pullman's power over an unpaid bill of $51. In 2018, APS cut off power to customers more than 110,000 times, according to the company's own data. Of those, more than 39,000 cutoffs happened during the particularly scorching months of May through September.

Pullman's death sparked wide media coverage and street protests over APS's disconnect policy and pushed Arizona regulators to ban power shutoffs on hot summer days. But Pullman's death also raises larger questions about the future of cities like Phoenix and Chennai in a rapidly warming world. As temperatures soar in the coming years, the real question is not whether superheated cities are sustainable. Phoenix is not going to melt into the desert and Chennai will not return to the jungle. The question is, sustainable for who? And at what cost? As cities grow and the heat rises, the future of Phoenix and Chennai, and many cities like them, is of a kind of temperature apartheid, where some people chill in a bubble of cool and others simmer in debilitating heat. This is not how you build a just, equitable, or peaceful world.

4. Life on the Run

In 2017, a few weeks after Hurricane Harvey turned the city of Houston into an aquarium, I drove across central Arizona on old Route 66, where the ghosts of the Dust Bowl, the 1930s drought that caused an area of the West the size of Pennsylvania to dry up and blow away, haunt every gas station and roadside ice cream shop. Novelist John Steinbeck called Route 66 the "Mother Road" — it was the route that hundreds of thousands of people took to escape from America's first man-made environmental catastrophe.

In Flagstaff, I turned south on Route 17, heading toward Phoenix, and watched the temperature gauge in the dashboard of my rental car climb to 107, 108, 109 degrees. The asphalt shimmered from the heat, and tinder-dry Ponderosa pines along the road looked like they were about to spontaneously combust. I pulled off the highway for gas and water. At the service station, I parked next to a Subaru with the words "We survived Hurricane Harvey, Orange Texas" scrawled on the back window in bright pink letters. The car was splattered with mud and dust, and luggage and boxes and a guitar case were piled in back.

A middle-aged woman and a scruffy man with wild brown hair pulled themselves out of the car. They looked road-weary and haggard. The man popped open the hood and fiddled with some wiring.

I nodded to the words on their back window. "Are you escaping Hurricane Harvey?"

"Yeah," the woman said. She introduced herself as Melanie Elliott. "We had to get out of there."

"It was bad," the man said, bent under the hood. His name was Andrew McGowan.

When Hurricane Harvey hit Houston, it dumped trillions of gallons of water onto a concrete-and-asphalt city that was not designed for climate catastrophe. Hurricanes are heat engines, powered by warm, moist air rising over warm oceans (that's why there are no hurricanes in the Arctic). The hotter our world gets, the more intense hurricanes become. During Harvey, hundreds of thousands of homes were inundated, and tens of thousands of people fled to hotels and shelters on higher ground. Others slept in their cars on overpasses.

"It was especially bad in Orange, where we lived," Elliott explained. Orange, I later learned, is an old industrial seaport town on the border with Louisiana, pop. 18,000. Orange has been hit repeatedly by recent hurricanes: In 2005, Rita ravaged the city. Three years later, Ike breeched the city's levee and flooded the streets with as much as fifteen feet of water. Three people died.

"We were just dealing with water all the time, constant flooding," McGowan said from under the hood. "The whole place is going under."

"Harvey was it for us," Elliott added. "Too much water, we can't deal with this anymore. We are going to San Diego."

"What are you going to do there?"

"We don't know," McGowan said, straightening up and closing the hood. "I'm gonna play some guitar and see what comes along."

"We just think there are opportunities there," Elliott said.

I wished them well, and they piled into their mud-splattered Subaru and headed toward the highway. I thought of the old Woody Guthrie song about the farmers fleeing the Dust Bowl: *We loaded up our jalopies and piled our families in, We rattled down that highway to never come back again.*

As the world heats up, it moves. This is true on the molecular level as well as on the species level. All creatures, from the ancient cedars of Lebanon to the microbes in the deep thermal vents at the bottom of the Pacific Ocean, have evolved within a basic temperature range, and if that range changes too much, they have to find a more habitable climate niche. For humans, the decision to stay or go from extremely hot places is often dependent on money, which buys access to cooling systems, clean water, and food. But most living things don't have the luxury of conditioned air or ordering a case of Pellegrino from Whole Foods. For them, adaptation often means moving to higher latitudes or higher elevations where it is cooler. If they can't find refuge, they die.

In the past decade, scientists who study the movement of animals have found that of the four thousand species that they'd tracked, between 40 and 70 percent had altered their distribution. On average, terrestrial creatures are moving nearly twenty kilometers every decade. Marine creatures, who are largely free of barriers to seeking cooler waters, are moving four times faster than land-based animals. Some animals are making spectacular leaps. Scientists estimate that Atlantic cod are moving north at a rate of a hundred miles per decade. In the Andes, frogs and fungi species have climbed four hundred meters upward over the past seventy years. Even the most seemingly immobile wild creatures are on the move. In Japan, several species of coral polyps, which form the branching thickets of coral reefs, were found by scientists to be traveling about two miles north every year.

Plants are on the move too. In the eastern US, trees are shifting north and westward at an average rate of about two miles per decade. Speed and direction vary for different types of trees: the conifers are mostly heading north, while broad-leafed and flowering trees, such as oaks and birches, move west. Among the speediest travelers that scientists know of are white spruce, which are migrating north at a rate of about sixty miles each decade.

Some plants and animals are more adaptable than others. Sharks can move from Florida to Maine, but mussels and urchins and starfish cannot quickly swim to cooler water. Polar bears need floating sea ice to hunt seals; without it, they starve. Bats are good at moving around and finding new habitats, as are many birds. But some are not.

One example: Thick-billed murres and snowy buntings, both common Arctic birds, are well adapted to the frozen north. But as the Arctic heats up, their dark feathers draw in the heat. Their body temperature can reach 115 degrees in the sun, even on a 75-degree day. It would be like working out in the peak of summer while wearing an insulated jacket. And because they are already living in one of the coldest places on the planet, migration to a cooler climate is out of the question.

Even animals that move freely have a rough time. Pacific salmon are cold-water species, adapted for mountain-fed, forested rivers and cool oceans and estuaries. Healthy water temperatures for salmon are under 58 degrees. Above 59 degrees, their bodies become stressed, making them easier targets for predators and at higher risk of disease. When water temperature hits between 70 and 72 degrees, it forms a "migration barrier," meaning it's too hot for salmon to swim through, and can be lethal. These extreme temperatures have a generational impact too. Even if heat-stressed salmon can survive to reach their spawning grounds, they are less successful in reproducing.

Warmer water also makes it tough for young hatchery-raised salmon to survive their trip from the hatchery to the sea. During the scorching summer of 2021, wildlife officials in California actually taxied young salmon through the hottest part of their journey in tanker trucks.

Coping with a warmer world has other risks too. On a trip to Botswana, I learned that lions prefer to hunt in the cooler parts of the day, mostly early morning and evening. That means that prey species, such as gemsbok, zebra, and waterbuck, switch to feeding at hotter times of the day when they are not in the presence of lions. But feeding during the heat of the day leaves them more vulnerable to heat stress — and less able to adapt to rising temperatures.

There is also the problem of synchronicity. Bumblebees are well suited to a colder climate, with long, furry hairs. But that makes them more vulnerable to even small changes in temperature. Bumblebee populations have been hardest hit in warming southern regions such as Spain and Mexico, where some species already live near the edge of their temperature range. Sometimes bumblebees get so hot that they fall dead out of the sky, said University of Ottawa biologist Peter Soroye. "Bumblebees are disappearing from areas eight times as fast as they are recolonizing others." And because bumblebees are important pollinators in wild landscapes (as well as for crops like tomatoes, squash, and berries), the decline of bumblebees ripples through ecosystems. Without a strong population of bumblebees in Montana, for example, the huckleberry bushes don't get pollinated. Which means they don't fruit. And because huckleberries are an important food for grizzlies, grizzlies have to look elsewhere for food — including dangerous journeys into the campsites of backcountry hikers.

Then there are the pine bark beetles, which are ravaging pine trees in the Rockies. When it comes to adapting to heat, insects have an advantage in that their lives are short and they can evolve quickly. Like all insects, pine bark beetles can't regulate their body temperatures. If it's cool, they are slow or dormant. As things warm up, they speed up. They eat more, reproduce more. And they don't die off in winter.

Pine bark beetles' favorite food is — surprise — pine bark. "Pine bark beetles are very much in tune with effects of climate on trees," Diana Six, an entomologist at the University of Montana, told me. "Beetles see the whole world chemically. They can actually ‘smell’ a tree that is stressed by heat and drought, and they go after those trees preferentially.” Recent beetle outbreaks in the West are roughly ten times larger than they were a decade ago as beetles sweep through heat and water-stressed forests of Colorado, Arizona, and Montana. They’re also moving into new regions of the West that had previously been too cold.

The beetle invasion not only kills vast swaths of forest, but also leaves the trees more vulnerable to fire. More wildfires lead to a cascade of consequences not just for the creatures who live in those forests (including humans), but also for the rest of us: injecting more carbon into the atmosphere, landing more soot on glaciers (which accelerates their melting because each tiny piece of black carbon is like a heat sink, absorbing sunlight far faster than the reflective ice of a glacier), and creating more burned-out open land. All this heats up the atmosphere even further. And that, in turn, amplifies the metabolism of bark beetles more, driving them to move to new forests, which they devour and which will, inevitably, burn. Scientists call this a positive feedback loop. And it’s evidence that the appetites of a few billion beetles can change the world.

■■■■

There were boxes all over Robert Stevens's apartment in North Phoenix. Some were sealed tight with packing tape, marked "CDs" or "Comix." Others were open, spilling out shirts or piled high with heavy computer programming textbooks. "I never realized how much shit I own," he muttered to himself. Stevens, twenty-nine, a slightly manic software programmer, was wearing jeans and a T-shirt and flip-flops. I'd met him twenty minutes earlier when I happened to drive by his apartment and noticed moving boxes by his dusty RAV4 and stopped to ask him where he was going. We chatted for a while, then he asked if I would give him a hand. "Let's get this done," he said as I helped him carry boxes out to the RAV4. The next morning, he was driving to Minneapolis, where he was moving in with his sister and planned to do some freelance coding.

"It's kinda beautiful here, I must admit," Stevens said, motioning to the jagged desert mountains out the window. He had grown up in Buffalo and followed his girlfriend to Phoenix four years ago. He loved the sunrises, and often got up early to go hike in the Phoenix Mountain Preserve. Hiking, in fact, was one of the things he thought he'd love most about Phoenix when he moved here. But as it turned out, it was on a hike where his romance with Arizona ended. "I was out on a trail last summer, and it was ridiculously hot, and I had gone too far, and I don't know — I just collapsed," he told me. "I totally fainted. Banged my head on a rock. Scared the hell out of my girlfriend. She gave me water, and I was okay, but it made me think — what am I doing living here? Maybe it's a genetic thing or whatever, but I can't take it. This heat is *dangerous.*"

Obviously, lots of people feel differently about Phoenix. Maricopa County, where Phoenix is located, had one of the highest population growth rates in the country in the 2020 census. People come for the jobs, the relatively inexpensive housing, and some, yes, for the weather.

"People have a lot of faith in adaptation," said UC Berkeley's Solomon Hsiang. "Whatever the temperature is, they assume they can deal with it. Just turn up the AC, you know? But when you're living it, you realize it's not so simple. And it costs a lot, too." As Hsiang pointed out, if people walked around in air-conditioned space suits, they would have no trouble with the searing heat in Arizona. "But who wants to live that way? And who can afford it?"

One of the central questions when it comes to adaptation is "Adaptation to what?" It's one thing to talk about three degrees of global warming over two or three decades. It's quite another to get hit by summer heat waves *right now* that are twenty or thirty degrees above what you expect. Sure, you could rebuild London so it's a nice place to live even if it were regularly hit by heat waves like the one that baked the city in 2022, but installing rail lines that don't melt and houses that aren't ovens and asphalt that doesn't turn to pudding would not only be hugely expensive, but would take decades to implement. In many places in the world today, heat is rising faster than our ability to adapt to it.

Some people may try to stick around and fight it out with Mother Nature, but most will not. "People will do what they have done for thousands of years," said Vivek Shandas of Portland State University. "They will migrate to better climates."

That's what happened during the Dust Bowl. Richard Hornbeck, a professor of economics at the University of Chicago who has studied the Dust Bowl extensively, argues that farmers could have adapted to changing conditions by switching to different plows, planting different crops, or shifting their fields to pastures for cattle or sheep. But they didn't. "There was too much inertia in how things have always been done, and too much investment in certain kinds of farm machinery, for people to make the changes needed," said Hornbeck. "Instead of adapting, many of them just headed to California."

Globally, the climate crisis has put people on the run. Increasingly unpredictable rainfall in Southeast Asia has made farming more difficult and has helped push more than eight million people to move toward the Middle East, Europe, and North America. In the African Sahel, millions of rural people are streaming toward the coasts and the cities amid drought and widespread crop failures. The UN estimates that four out of five African countries don't have sustainably managed water resources and that seven hundred million people will be on the move by 2030. In 2022, catastrophic floods in Pakistan — caused in part by the heat-driven melt of glaciers in the Himalayas, and in part by the fact that hotter air holds more water — displaced 33 million people, which is about 15 percent of the entire Pakistani population.

"Should the flight away from hot climates reach the scale that current research suggests is likely," writes journalist Abrahm Lustgarten, "it will amount to a vast remapping of the world's populations."

The political consequences of these massive shifts of people are impossible to overstate. Here in the US, unfounded fears of brown people invading the country and stealing jobs and committing crimes are disguised as a policy debate about immigration. Fear of outsiders also fuels the rise of extreme right-wing politics in Europe and Australia. ("If you want to understand Australian politics," one Australian entrepreneur told me during my visit to Melbourne, "the first thing you have to understand is our fear of yellow hordes from the north.") Migration is driven by many factors, but a lack of food and water, both of which are exacerbated by extreme heat, is pretty high on the list.

One of the big surprises in the 2020 US census, however, is the degree to which Americans are moving *toward* places with the highest risk of climate impacts — especially extreme heat. According to an analysis by Redfin, a real estate brokerage firm, the fifty US counties with the largest share of homes facing high heat risk saw their populations increase by an average of 4.7 percent from 2016 through 2020 due to migration. In contrast, the counties with the largest percentage of homes facing high storm risk experienced average population growth of only 0.4 percent.

How to explain this? Denial and ignorance are one answer, especially among politicians whose job it is to create policy that discourages people from moving into harm's way. As the US Government Accountability Office concluded in a 2020 report, "Unclear federal leadership is the key challenge to climate migration as a resilience strategy."

Also, places with higher climate risks tend to be cheaper to live in and have more space and access to nature (for the moment, anyway). According to the Redfin analysis, storms (including winter storms) are the only climate risk people are moving away from, with counties with the smallest share of risk growing faster than the counties with the largest share of risk. So when you look at migration in the US, it's fair to say that people are moving away from storms and toward heat.

According to Redfin's analysis, the most attractive hot place in America is Williamson County, Texas, which is part of the Austin metro area. Williamson County, where every home faces high heat risk, saw its population grow 16 percent due to positive net migration from 2016 through 2020. That's the highest growth rate of the fifty counties the company analyzed. It was followed by Pinal County, Arizona (just south of Phoenix), which had a positive net migration rate of 15 percent. Rounding out the top five were three Florida counties: Pasco County, Osceola County, and Manatee County.

Williamson County has exploded in recent years with malls, condos, taco bars, and office buildings. Dell, Amazon, and Apple have big new campuses in the area, and Tesla's headquarters and factory, which employs ten thousand people and is three times as large as the Pentagon, is only a few miles away.

Climate-wise, Texas is the belly of the beast. "If you live in Kansas, you don't really have to worry about hurricanes or wildfires, and if you live in Oregon, you have to worry about wildfires and drought, not worry about hurricanes," says Texas climate scientist Katharine Hayhoe. "If you live in Texas, you have to worry about everything."

As it turns out, I have some insight into why so many people are moving to hot places, because I did it myself. For twenty years, I lived in upstate New York, which is one of the best places to live in America if you take climate change seriously. It's not too hot in the summer, there is plenty of farmland and water, it's far from the rising seas, it has a politically engaged local population, easy access to the Adirondacks, a strong arts culture, and decent transportation, including one of the great train rides in America down the Hudson to New York City. So why leave?

Because I fell in love. One day I was at the racetrack in Saratoga Springs and met a woman in a yellow dress named Simone. As it turned out, she lived and worked in Austin, and had decided she needed to get out of Texas for a few days in late August because it was just too damn hot. So she flew to upstate New York, where she knew it would be cooler, to visit a friend — and my life changed. Within a year, I had left the green hills of upstate New York and was living in one of the hottest cities in America. Or to put it another way, because Simone decided to flee the heat for the weekend, I ended up moving toward it.

If nothing else, my move shows the complexity of climate-related decisions about where and how we live. I knew very well that my move made no sense from a climate perspective, but who cares? I would rather live with Simone in a tent in the Sahara than without her in a solar-powered house on a lake in the mountains. Austin is full of people moving here for similarly personal reasons, whether it is the music scene or a high-tech job. It has created traffic jams and insanely inflated real estate prices, but that's how it goes.

While I was researching a previous book I wrote about sea-level rise, I met dozens of people who were anguished about whether or not to sell their homes near the beach. They all loved the ocean, loved their friends, loved their community. For some, the desire to move was driven by a fear that as the seas rose, the value of their homes would decline. For others, it was rising insurance costs or concerns about physical risk from a big hurricane. It was never a simple or easy calculation.

For a middle-class American, the risk of extreme heat seems, at first glance, much more manageable. Summer days can be suffocatingly hot here in Texas. But if it gets too hot, you stay indoors with the air-conditioning on. You run errands in the early morning. You schedule outdoor projects for fall or spring. You wear hats and cool clothes. You drink a lot of water. And you wait for the heat to pass.

Sometimes life here seems tenuous. That

might sound like a weird thing to say about a booming city like Austin, but it's true. Since I moved to Texas, I'm very aware that my life here is dependent on technology — not just the air-conditioning itself, but the grid that powers it, as well as the long, complex chain of political and economic logic that keeps the grid up and running. Like millions of other Texans, I got a lesson in how fragile that system is in the winter of 2021, when the power went out in the middle of a freakish ice storm. Suddenly, Simone and I were huddled in the darkness, freezing, unable to drive anywhere because the streets were ice sheets. It was a reminder that the easy comfort of modern life is not so easy at all. That goes for extreme heat too. When the temperature climbs over 100 degrees and I've taken refuge in our air-conditioned house, the heat feels like a Sword of Damocles hanging above the city by an electric wire.

El Camino del Diablo, or the Devil's Highway, is an ancient passage in the Sonoran Desert of Arizona that runs through Organ Pipe Cactus National Monument, a US Air Force bombing range, and the Tohono O'odham Nation reservation, and more or less directly over the grave of writer and desert hell-raiser

Edward Abbey.* The Devil's Highway is a haunted place. Stories of people struggling in the desert's merciless heat, tongues swollen from lack of water, stumbling, hallucinating, stripping off their clothes, go back centuries. Here among the saguaro cactus and ironwood trees, suffering from heat is an ancient and tragic ritual. Here, heat is not so much an engine of migration as a migration barrier, a thermal wall that blocks or kills anything that attempts to cross it, just as warm water in a river is a migration barrier for spawning salmon in the Pacific Northwest.

One person who understands all this as well as anyone is John Orlowski, a member of No More Deaths, a humanitarian group that helps migrants in their dangerous journey across the US/Mexican border. Orlowski is in his early sixties, with a shock of white hair and a deep desert tan. He has the strong, lanky build of a lifelong mountain climber (he has scaled Yosemite's El Capitan three times), with a nose that tilts off to the right like it was broken in a fight back in the day. While other men his age settle into retirement in a condo in Florida or the Rockies,

* That's the rumor, anyway. Officially, the site of Abbey's grave has never been disclosed.

Orlowski moved to Ajo (pop. 3,600), an old copper mining town on the Mexican border about two hours from Tucson, to help migrants get safely across the desert.

I met Orlowski one morning for breakfast at the Agave Grill, a restaurant in Ajo decorated with watercolors of saguaro cactuses and rattlesnakes. "Ajo is a heavily militarized place," Orlowski told me. Besides the fifteen sheriffs in this sparsely populated county, there is a new US Border Patrol station just outside of town, where five hundred Border Patrol agents work. Still, the desert around here is vast, and the migrants are many. Over coffee, Orlowski explained how the border police use helicopters to patrol the most remote parts of the borderlands. If they spot a group of migrants, they use a technique called dusting, in which they lower the helicopter down to thirty or forty feet above the migrants, kicking up a huge dust storm and scattering the migrants. "When they are alone, they are much more vulnerable," Orlowski said. It is a brutal technique. Many of the people who die have gotten separated from their families and travel companions and end up wandering alone in the desert. In the last three decades, No More Deaths estimates that more than nine thousand people have died here on the

Devil's Highway, virtually all of them from dehydration and heat exhaustion.

Orlowski pointed out that there are lots of Border Patrol agents in places where it is easy to cross. And far, far fewer in places where it is hot and dangerous to cross. "Part of their strategy is to funnel migrants through the hottest, most dangerous regions of the border," Orlowski explained.

"So, basically, the US Border Patrol has figured out a way to weaponize heat," I said.

"Yes, that's one way to think about it," he replied.

After we finished breakfast, we got into his dusty pickup and headed out into the desert. In the back, he was carrying plastic jugs of water, cans of beans, and other food items that can survive for weeks in the desert. We drove through Organ Pipe Cactus National Monument, where saguaro cacti stand like worshippers with arms raised to the heavens. The rocky peaks of Growler Mountains, harsh and mean-looking, watched over us in the distance. We eventually turned down a dirt road and drove parallel with the border. In some places, the border itself is just a 4x4 fence that you could easily walk over or under. In other places, a tall, grotesquely ugly metal barrier (Trump's wall) had gone up. Every once in a while, I'd see a white cross, which marked the spot where human remains had been found.

After about forty-five minutes of bouncing along the dirt road, we pulled off and parked. It felt like we were a thousand miles from anything soft, cool, or kind. We loaded jugs of water into backpacks and headed out for a spot on the top of a nearby mountain where Orlowski knew migrants cross. I had prepared for the hike pretty well, I thought, with a wide-brimmed hat, sunglasses, and pants and a long-sleeve shirt with UVF protection. Still, I felt the heat immediately and began to sweat. I saw more white crosses. As we plodded along, I worried about stepping on human remains that had been turned to near dust by the heat. A few days earlier, the body of Gurupreet Kaur, a six-year-old girl from India who was crossing the border with her mother and other family members, was discovered within a few hundred yards from where we were hiking. We found lots of evidence that people had been through here recently: a worn Nike running shoe, a shirt, a plastic bag, a phone charging cord, a few black plastic one-gallon water bottles (they are black so they don't reflect light and the Border Patrol doesn't see them from far away). As we headed up the mountain, I couldn't help but notice the austere beauty of the desert: red flowers of the prickly pear cactus in bloom, and spiny ocotillos, their long, slender branches like octopus tentacles.

After about an hour of hiking, we reached the mountaintop. Orlowski and his fellow volunteers at No More Deaths had dropped water and beans here about a week earlier. Now the supplies were gone — a sure sign that migrants had come through. We pulled six gallon jugs of water out of our backpacks. *"Agua Pura,"* Orlowski scribbled on the jugs with a marker. He unloaded an eight-pack of SunVista pinto beans out of his backpack and left that too. In the distance, I saw towers of the US Border Patrol's alert beacons. A helicopter skimmed by. We could see south to Mexico, and north toward Tucson and Phoenix. I was exhausted. I was hot. I sat on a rock beside Orlowski and tried to imagine wanting to come to America so badly that I would walk for five or six days across this ghostly boneyard of heat. And the hotter it gets, the more treacherous this passage will become. Migration itself becomes a deadly gamble.

Orlowski pointed toward the Growler Mountains in the distance. "Between here and there, I'm sure there are dozens of people crossing right now, you just can't see them," he said. "Just like the heat, they are invisible."

5. Anatomy of a Crime Scene

Early in the afternoon on July 19, 2022, the hottest day in recorded history in the UK, thirty-nine-year-old climate scientist Friederike Otto rode her bike across London Bridge from her flat in the Southwark borough to a meeting downtown. She is thin, with green eyes and a shy smile. Tiny silver daggers dangled in her pierced ears and colorful beaded bracelets dangled on her wrists. She is the kind of person who wears green Converse high-tops everywhere and is almost as serious about modern dance as she is about modern climate (once, when asked by a journalist to describe herself with four words, Otto said: "Physicist, wannabe-dancer, media-go-to-scientist, theatreland"). Her twelve-year-old son was off at school, where — to her relief — she knew the administration had taken precautions against the heat, making sure students stayed inside and had plenty of water to drink.

When the heat arrived the day before, Otto was prepared. Her 1842 Victorian house didn't have air-conditioning — because of London's historically mild climate, almost no one in the city does — but it was insulated. It also had a basement, which was cool, where her dog, Skyler, a gangly Lab-collie mix, sought refuge when the heat started to rise. To Otto, who spent most of her waking life thinking about the causes and consequences of extreme weather events, living through a heat wave was like being trapped in her own imagination. It was both familiar and surreal.

As she pedaled over the Thames, the hot wind blowing over the river burned her lungs. It surprised her. And her surprise surprised her. After all, she had been studying heat waves for more than a decade, and is a leader in a revolutionary new climate science called extreme event attribution, which has pioneered new ways of determining how rising levels of CO_2 in the atmosphere are changing the frequency and intensity of extreme weather events. Otto and her colleagues have proven that almost all extreme heat waves we are experiencing today are not the standard effects of Mother Nature, or, as climate skeptics like to say, "just the weather." They are caused by the actions of human beings and the choices they make.

The hot wind spooked Otto. After she crossed the bridge, she rode by the monument to the Great Fire of London, which burned most of the city down in 1666. To Otto, the 202-foot-tall Doric column with a gilded urn of fire at the top was a taunting reminder of what was at stake. "You could feel the dryness in the hot air and that hot wind," Otto told me later. "All it would take is one spark, and I could imagine London on fire again."*

She pedaled on through the city. She was surprised to see how many people had their windows open — which, Otto knew, might seem logical but is exactly the wrong thing to do when you don't have air-conditioning (in the city, especially if there are no cooling winds, it's better to close the windows and draw the curtains early in the day to keep out both the sun and the heat). As Otto biked along, she knew people were dying in the heat at that very moment. People trapped in hot flats. People with heart

* In fact, parts of London did burn. Grass fires spread to homes, shops, and vehicles in what city officials described as the Fire Brigade's busiest day since World War II.

conditions. People who were working outside and, because they live in a place where heat has been a stranger, have no idea what to do when the temperature spikes. To Otto, she wasn't just a scientist biking through London on a hot day. She was a detective biking through a crime scene.

To understand why heat waves are so dangerous, it helps to understand a little bit about the nature of heat itself. After all, heat is not the same thing as temperature. Temperature is a *measure* of heat. But what is heat? A chemical reaction? A fundamental force like gravity? Pulsing electromagnetic waves like electricity? When you touch a child's forehead and it feels warm, or you grab the too-hot handle of a pan, what are you actually feeling?

The usefulness of heat as a tool has been well understood for thousands of years. Fire was domesticated not only as a means to cook and preserve food, but also to allow adventurous humans to move into cooler climates and keep warm. The Egyptians used heat to smelt copper and tin into bronze, heralding a new age of art and war. In Hinduism, heat is a path to enlightenment. The Aztecs painted their gods in red ochre, symbolizing what one scholar described as "their privileged relationship with heat." For many Native Americans, sweat lodges are linked with spiritual rituals, such as the Sun Dance of the Lakota Sioux, in which participants sought to reconnect with nature and the supernatural.

The ancient Greeks believed the world was made up of four elements: fire, water, earth, and air. One substance could be changed into another by altering the relative proportions of the four elements. Heating clay in an oven, for example, could be thought of as driving off water and adding fire, thereby transforming the clay into a pot. A few Greek philosophers, including Plato, rightly suspected there was a link between heat and motion, but didn't take it any further than that: "heat and fire . . . are themselves begotten by impact and friction: but this is motion. Are not these the origins of fire?"

Fifteen hundred years later, during the Islamic Golden Age in the tenth and eleventh centuries, a pair of thinkers proposed that heat was related to light — which was a big step in the right direction. Al-Bīrunī, a mathematician and scholar who served the Sultan Mahmūd of the Ghazna Kingdom (now Afghanistan), was the first to divide hours into minutes and seconds. He was also the first to ask how heat got from the sun to the Earth. His answer: "Heat occurs here from the reflection of light." Al-Bīrunī eventually decided that heat is nothing more than "detached rays of the sun" but also noted that it was caused by "friction." This fascination with light and heat was shared by Ibn al-Haytham, an Arab astronomer and mathematician who was one of the founders of the study of optics. He used mirrors to experiment with directing and concentrating light, and came to the conclusion that light and heat were connected — increasing the concentration of light would heat objects more quickly.

The first attempt to measure heat in a scientifically quantifiable way came with the invention of the thermometer. Around 1602, Galileo built a device for measuring temperature that depended on the expansion of air when it was heated. It was basically a glass tube filled with liquid (wine was sometimes used), in which smaller sealed glass spheres floated. As the temperature rose, the air in the spheres expanded, causing the spheres to rise in the glass tube.

The modern thermometer was born a hundred years later, when German physicist G. D. Fahrenheit built one with alcohol in a bulb at one end of a thin uniform tube that was evacuated and sealed at the other end. The alcohol (later replaced by mercury) moved up and down in the tube as the temperature changed. The Swedish astronomer Anders Celsius made a similar thermometer in 1742. Their thermometers were graduated in different scales, which subsequently became known as Fahrenheit and Centigrade (or Celsius) scales.*

Thermometers were useful, but they didn't help much in advancing any deeper understanding of what heat actually *is*. The first person to really grasp the cosmic nature of heat was a tall kid from Massachusetts who was known for most of his adult life as Count Rumford. Born on a farm in 1753, Rumford led an outlandishly colorful life — first as a spy for the British army during the Revolutionary War, then in the court of Karl Theodor, Elector of Bavaria, in the decrepit and declining Holy Roman Empire. He lived

* On the Fahrenheit scale, the freezing point of water at sea level is 32 degrees and the boiling point is 212 degrees. Celsius measured the freezing and boiling points of water and divided the distance between the two into hundredths. He originally designed the scale in the opposite order of the scale used today — 0°C was the boiling point of water, and 100°C was the freezing point — but other scientists later reversed the scale.

in a gilt-walled palace, partied with Mozart and Voltaire, and, according to one giddy biographer, slept with half the countesses and duchesses on the Continent. He also happened to be a brilliant, intuitive thinker. Rumford's contribution to the science of heat gave birth to conceptual breakthroughs that led to the first and second laws of thermodynamics, two fundamental laws of physics that underpin our understanding of the universe and, not coincidentally, are key drivers in the heat waves that are killing people today.

During Rumford's time, there were two competing theories of heat. One was the kinetic theory. The basic idea was that the heat of a body was associated with the constant movement of the particles of which the body was made. Frictional rubbing or hammering, such as a wheel spinning on an axle, increased this movement so that the body got hotter. As Isaac Newton put it in the early eighteenth century, a few decades prior to Rumford's experiments: "Heat consists in the trembling agitation of the smallest parts of bodies all manner of ways; & the parts of all bodies are always in some agitation."

But the kinetic theory of heat had been pushed aside by the caloric theory of heat, which came to the fore a few decades later in the late eighteenth century. It owed much of its popularity to French chemist Antoine Lavoisier, one of the scientific rock stars of the late eighteenth century, a man who is often referred to today as the father of modern chemistry.

The caloric theory supposed that heat was an invisible substance or fluid that somehow flowed into a body when it was heated and flowed out when it was cooled. The hotter a body became, the more caloric it contained, which explained why a body's volume increased (think of how your fingers swell up on a hot day). Lavoisier regarded caloric as a real substance and even included it, along with light, in his table of thirty-three chemical elements, which he produced in 1789.

Rumford thought the caloric theory was nonsense. He also knew that if he could prove it, he would be celebrated and famous.

In 1797, some ten years after Lavoisier produced his table of elements, Rumford, as major general and commandant in the court of the Holy Roman Emperor of Bavaria, decided it was time to upgrade Bavaria's artillery for the defense of Munich. And for that purpose he ordered the manufacture of heavy brass cannons.

This was not a neat and tidy project. Rumford's shop was, in fact, positively Frankenstein-ish, with dim light and tiny soot-covered windows and screeching metal biting into metal and the acrid smell of freshly cut brass. As one science historian notes, "the immediate purpose of making cannons was not to satisfy scientific curiosity but to kill Frenchmen."

Here is how it worked: Each cannon barrel, after being cast in the form of a solid cylinder, was placed horizontally on the bed of a lathe. A massive screw pushed the stationary drill bit — a blade made of hardened steel — into the front end of the barrel with a force of several tons. At the same time, a revolving shaft attached to the back of the barrel turned the entire casting on its own axis at the rate of thirty-two revolutions per minute. The power for turning the cannon was provided by two draft horses harnessed to a winch. As they walked in a circle, a system of gears transmitted the motion of the winch up to the lathe's shaft.

At some point (Rumford's journals are not clear about exactly when), he discovered that the cannon-boring apparatus could be modified into a heat experiment. He built a tub that allowed him to submerge the cannon in water while it was being drilled. A tub of water, Rumford knew, was a better reservoir for heat than the air, and one that he could measure precisely. It was also a result that you didn't have to be a scientist to understand: boiling water is the most obvious and everyday application of heat.

Rumford reasoned that if heat was an invisible fluid, or a substance contained in the material of the brass cannon itself, as the caloric theory of heat suggested, then surely that substance was not infinite. It might indeed cause the water to heat up, but for how long? With the cannon-boring apparatus, Rumford could keep the experiment going as long as the horses kept walking. Although he didn't think about it this way at the time, he was transforming the energy of the horses into mechanical energy in the spinning of the drill bit, which was then converted into thermal energy by warming the water. It was a perfect illustration of the first law of thermodynamics, which would not be understood for another fifty years or so, but which says that energy can be transformed from one form to another, but it cannot be created or destroyed.

On a gloomy day in October 1797, Rumford's experiment was ready to go. He commanded the horses to begin walking. The gears spun, the drill bit pressed against the blank cannon. Rumford kept his eye on the thermometer in the tank: After 60 minutes, the temperature of the water had risen from 60 degrees to 107 degrees; after 90 minutes, to 150 degrees; after 120 minutes, to 177 degrees; and after 150 minutes, the water actually boiled.

"It would be difficult to describe," Rumford later wrote, "the surprise and the astonishment expressed in the countenances of the bystanders, on seeing so large a quantity of cold water heated, and actually made to boil, without any fire."

In the history of science, this was a revolutionary moment. Over the next few decades, German physicist Rudolf Clausius and British mathematician Lord Kelvin built on Rumford's work and established a dynamic theory of heat that underlies modern physics and chemistry today. But with this single experiment, Rumford proved that heat is not an invisible fluid but an inexhaustible expression of matter, something built into our world not unlike the way language is built into our lives. There is no end to the words you can say, just as there is no end to the heat that can be generated. He didn't know it at the time, but he had found the link between the sun and the horses and his own life and everything around him. They were all manifestations of energy — which is to say, of motion, and of heat.

Thanks to Rumford, and those who followed in his footsteps, it is now pretty easy to say what heat is: it's the vibration of molecules. Or to put it another way, temperature is the average speed of a collection of molecules. Something is cold when the average speed of its molecules is low and it is hot when the average speed is high.

Still, heat does flow — but it doesn't flow like a river. Grab hold of the door handle of your car on a hot summer day and it feels like heat is moving into your hand. But that's not what is happening. When you grab the door handle, its faster-moving molecules bump into the slower-moving molecules in your hand, which causes the speed of the molecules in your hand to increase and those in the handle to slow down. Your brain senses the increased speed of the molecules in your hand as warmth; the temperature of your hand has increased. But there is no actual current of heat flowing into your hand. The molecules in the door handle stay in the door handle, and the molecules in your hand stay in your hand. The molecules in the door handle are just agitated, and they spread that agitation into your hand. "Heat moves less like a river through a canyon and more like laughter through a crowd," writes physicist Brian Greene.

Different molecules vibrate at different rates, depending on their structure. That is one reason why water holds more heat than air. Or steel more than wood. It is also why some gases trap heat in the atmosphere while other gases don't. Hydrogen, for example, is not a greenhouse gas, but CO_2 is. What's the difference? CO_2, like other greenhouse gases, has a molecular structure that makes it more sensitive to the heat that the Earth absorbs from the sun and reflects back into the atmosphere, creating thermal radiation. CO_2 molecules react to this radiation, causing them to vibrate and bend and jitterbug. As more greenhouse gases accumulate in the atmosphere, the jitterbugging in the sky above us accelerates. That's why, as we burn fossil fuels and release CO_2 molecules into the atmosphere, our world is heating up. The sky is literally vibrating faster.

Early pioneers of what we now call climate science, such as Eunice Newton Foote, whose 1856 paper "On the Heat in the Sun's Rays" was the first scientific article to state that if CO_2 levels were higher, the planet would be warmer, knew nothing of vibrating molecules. She and other researchers (they didn't call themselves scientists then) just understood that certain gases, such as CO_2, were good at trapping heat. However, Foote and other researchers certainly weren't worried about heat waves. Those who thought about the effect on the planet at all believed that a little warming would be a nice thing to have. Svante Arrhenius, who published the first calculations that described the greenhouse effect in 1896, claimed that the world "may hope to enjoy ages with more equable and better climates." Most people assumed that a "balance of nature" made catastrophic consequences impossible, and if any change did result from the "progress" of human industry, it would be all to the good. "Nobody worried about the impacts of a climate change that scientists expected would only affect their remote descendants, several centuries in the future, if it happened at all," science historian Spencer Weart writes.

Fast forward fifty years. In the 1950s, a few scientists realized that the level of CO_2 in the atmosphere might be rising, suggesting that the average global temperature might climb a few degrees before the end of the twenty-first century. One prominent scientist, Roger Revelle, speculated that in the twenty-first century the greenhouse effect might in fact exert "a violent effect on the earth's climate." In 1957, Revelle told a Congressional committee that the greenhouse effect might someday turn Southern California and Texas into "real deserts."

Among scientists, most of the worries about global warming, to the degree that there were any, were around melting ice. That was easy to understand: it gets hot, ice melts.

Then in 1984, a landmark paper on heat waves by scientists Linda Mearns and Stephen Schneider was one of the first to explore the possibility of extreme heat events and their connection to things we care about. It wasn't just that the world was getting warmer, they pointed out. "Even small changes in [average] temperatures sometimes result in large changes in event probabilities," they wrote. As an example, they showed that a five-day heat wave of at least 95 degrees in Iowa — which could be devastating to corn crops — was three times more likely with an average global warming of 3 degrees.

An even bigger turning point was the 1988 Congressional testimony of NASA scientist James Hansen, the man who is now thought of as the godfather of modern climate science. "I would like to draw three main conclusions," Hansen told a packed Senate chamber. "Number one, the earth is warmer in 1988 than at any time in the history of instrumental measurements. Number two, the global warming is now large enough that we can ascribe with a high degree of confidence a cause-and-effect relationship to the greenhouse effect. And number three, our computer climate simulations indicate that the greenhouse effect is already large enough to begin to affect the probability of extreme events such as summer heat waves."

Hansen showed a graph of global temperature over the last hundred years, pointing out that the rate of warming in the past twenty-five years was the highest on record. He detected warming of 0.4°C by 1988. The probability of a chance warming of that magnitude is about 1 percent. "In my opinion," Hansen famously stated, "the greenhouse effect has been detected, and it is changing our climate now."

Not surprisingly, Hansen focused on heat waves as the most obvious manifestation of rising temperatures. He compared the frequency of hot summers in Washington, DC, and Omaha, Nebraska. Between 1950 and 1980, Hansen argued, the probability of having a hot summer was 33 percent. But by the 1990s, because of the greenhouse effect, the probability increased to between 55 percent and 70 percent.

When it came to mechanisms for how that works, beyond the heat-trapping effect of greenhouse gases, Hansen was vague. He showed some model images of the difference between temperatures in the US now and in 2029, predicting that by 2029, it would be warmer almost everywhere. He pointed out that he noticed "a clear tendency" in his model for greater-than-average warming in the Southeast and the Midwest. "In our model this result seems to arise because the Atlantic Ocean off the coast of the United States warms more slowly than the land," Hansen said. "This leads to high pressure along the east coast and circulation of warm air north into the Midwest or Southeast."

But then Hansen issued his caveats: "There is only a tendency for this phenomenon. It is certainly not going to happen every time, and climate models are certainly imperfect tools. However, we conclude that there is evidence that the greenhouse effect increases the likelihood of heat wave drought situations in the Southeast and Midwest United States even though we cannot blame a specific heat wave or drought on the greenhouse effect."

This was a crucial point, one that was true of the science at the time, but one that would long be exploited by climate deniers and others. Climate change might change the odds of having a heat wave, they would argue, but you couldn't point to any one heat wave and say, "Climate change caused that." And Hansen was of course right — for a long time, you couldn't.

But thanks to the work of Otto and others, now you can.

When Hansen testified before Congress, Otto was six years old. She was born in 1982 in Kiel, Germany. She grew up in a cottage in the countryside not far from the Danish border, a place she called "excruciatingly boring." Her father was a university administrator; her mother taught English and Russian. Otto was a bookish, shy kid. "I was not a great student," she told me. She had no idea what she wanted to do with her life. She enrolled at the University of Potsdam, where she studied physics because it was the least-bad option among studies available to her. During the 2003 heat wave in Europe, which killed seventy thousand people — that's more than the number of US soldiers killed during the entire Vietnam War — Otto was in her first year of university. "Climate change was definitely not a topic — at least not in my world," Otto recalled. "I used to work in an ice cream shop during the summer. I remember that I regretted it that year because it was so busy that summer. The job usually had a lot of quiet time when I could read, but in the summer of 2003 it was extremely stressful because there were so many people wanting to buy ice cream."

By the time Otto graduated in 2007, she began hearing and reading about climate change. It interested her. "I talked at that time to my dad about maybe going into climate science and he thought it a terrible idea," she told me. "He thought it's just a fashionable topic nobody will be interested in in a few years' time."

At university, she discovered she liked reading about science almost as much as she liked the science itself. So she moved to Berlin to attend the Free University of Berlin and pursue a PhD in the philosophy of science. "I got interested in models — social models, economic models, climate models. I was interested in what can be known and what can't." She eventually focused her attention on climate models — not because she was concerned about saving the world, but because how you model a system as infinitely complex as the Earth interested her. Her 2011 PhD dissertation was titled: "Modelling the Earth's Climate — an

Epistemic Perspective." It's a 124-page argument about how to reduce uncertainty and increase transparency so that models are more reliable and trustworthy. And getting models right is important, Otto argues, because scientists can't run experiments on the Earth's atmosphere itself. "We have only got one Earth," she wrote.

After she finished her dissertation, Otto had no idea what to do next. Who wants to hire a PhD in philosophy of science? "I was completely unemployable," she says. But then, she applied for a short-term position with Myles Allen, a geophysicist at the University of Oxford. Allen was interested in how to make climate models better too. Otto got the job and moved to Oxford. It was a lucky break.

One of the most consequential revolutions in climate science began in 2003, when Allen watched the flooding River Thames rise closer and closer to the walls of his home in South Oxford. He thought, *I'm a climate scientist. Why can't I find out who is responsible for this?*

As Allen monitored the rising waters, he happened to hear someone from the Met Office, the UK's national weather service, on the radio in his house telling him that discovering who was responsible was impossible. Sure, the flood was the *type* of event likely to be made more frequent by global warming, the voice on the radio said. But saying anything more concrete about the cause-and-effect relationship was not possible. It was the same point Hansen had made twenty years earlier. But was it really true any longer?

Allen, who is known as a fiercely independent thinker, took it as a challenge to find out.

In an article he wrote as the floodwaters inched closer to his kitchen door, Allen argued that it might not *always* be impossible to attribute extreme weather events to climate change — it was just "simply impossible at present, given our current state of understanding of the climate system." And if researchers were ever able to make that breakthrough, he mused, the science could potentially influence the public's ability to blame greenhouse gas emitters for the damages caused by climate-related events. Allen's commentary was published in February 2003 in the prestigious journal *Nature* with the stark but scary (for fossil fuel companies) title "Liability for Climate Change."

Allen got a chance to experiment with his idea a few months later when a heat wave baked Europe with the highest temperatures since records began being kept in the fifteenth century. Allen teamed up with Peter Stott, a climate scientist at the Met Office, to explore whether the heat wave could be linked to climate change. The details are complex, but their basic method was to use climate models with lower concentrations of CO_2 in the atmosphere to see if they would produce a similar heat wave. Their conclusion: "We estimate it is very likely that human influence has at least doubled the risk of a heat wave exceeding this threshold magnitude," they wrote in a paper published the following year.

With that sentence, the science of extreme event attribution was born.

There were a lot of skeptics, including scientists who questioned their methods or their data. Others argued that the models didn't accurately represent the real world. But Allen and Stott were highly respected scientists. Their findings were not easy to brush away.

It's no surprise Allen and Stott completed their first attribution study on a heat wave instead of, say, a hurricane. "Climate change is reflected more strongly in heat waves than in any other phenomena," Otto said. A lot of variables go into the intensity and duration of a heat wave, including the moisture in the land and atmospheric circulation patterns. But compared to the complex dynamics of a hurricane, they are fairly easy to define and model.

When Otto arrived in Oxford in 2011 to work with Allen, she had never heard of attribution studies. But she was intrigued. And the questions these kinds of studies raised dovetailed with a lot of what she had thought and written in her PhD dissertation.

She went to work immediately. In 2010, a major heat wave cooked Russia with temperatures as high as 104 degrees. More than 55,000 people died. She wondered, could this extreme event be directly attributed to climate change?

At first, the answer wasn't clear. One paper published shortly after the event by scientists from Boulder, Colorado, concluded that the heat wave was "mainly due to natural internal atmospheric variability." In another study, however, scientists from Potsdam, Germany, calculated an 80 percent probability that these record temperatures would not have occurred without climate change. Otto and colleagues, including Dutch climatologist Geert Jan van Oldenborgh, spent a few weeks analyzing whether the studies really did contradict one another — and if so, which one was correct. "The result was a pleasant surprise for all involved," Otto wrote later in her book, *Angry Weather*. "Both studies were correct; they had simply asked different questions. One study focused on the heat record itself—the magnitude of the heat wave — while the other looked at the probability of the heat record being broken. So had climate change made the heat wave more likely? The answer was a resounding 'yes.' "

Otto and her colleagues' review of the two studies — it was her first published paper ever — helped cement extreme event attribution as a legitimate branch of climate science. Her work was even noted in the Fifth Assessment Report of the United Nations' Intergovernmental Panel on Climate Change, the gold standard of climate science, which was completed in 2014.

The science of extreme event attribution, Otto quickly realized, could have a profound effect on how people viewed the impacts of human-made climate change — and whom they held responsible for it. "To me, science is — or can be — a tool for justice," she told me. "Extreme event attribution is the first science ever developed with the court in mind."

For the next few years, Otto refined her

methods for connecting the dots between extreme weather and climate change. Some events, Otto and her colleagues discovered, such as a four-day rainfall in Europe that caused severe flooding in the Danube and Elbe Rivers, was not linked to climate change. To Otto, saying that some events were not related to human-made climate change was as important as saying that some were. But either way, none of her work had much impact, in part because her papers came out so long after the event that nobody paid attention.

In 2014, Otto and Allen happened to be at a scientific conference in San Francisco. In a Starbucks downtown, they met with Heidi Cullen, the chief scientist for Climate Central, whose mission it was to make science more accessible to a general audience. Cullen understood the power of extreme event attribution, but thought that if it was really going to have an impact on how people viewed climate change, the link between extreme weather and climate had to happen faster, in real time, while the event was still fresh in people's minds. "Can you do this faster?" she asked. Otto and Allen thought it was possible. But working faster meant risking mistakes in their analysis, which could set back climate science for years. Still, Otto was game to try (Allen was focused on other work). A few months later, she and van Oldenborgh started World Weather Attribution, a project to bring together scientists devoted to rapid analysis of extreme weather events to determine how (and if) they were caused by climate change. "It was as though, two years after the light bulb was first developed, we had announced that we were going to install electric lanterns on every street without knowing whether mass production was even possible," Otto later wrote.

More than a few people in the climate science establishment believed that real-time attribution wasn't "real" science, in part because each paper didn't go through the yearlong peer review process that other papers do before publication (Otto makes clear that the models she uses are peer-reviewed, just not the results themselves). "I think the pushback would have been much greater if I weren't a young woman," Otto explained. "I think a lot of the old white guys in the science world just dismissed me as a crazy woman off doing something on her own, and paid no attention to me."

Not every event Otto and her team studied was amplified by climate change. They analyzed Hurricane Harvey, which hit Texas in 2017, and determined that climate change made the storm three to four times more likely. They found that the flooding in Bangladesh the same year, however, was *not* linked to climate change. By the time the heat wave hit the Pacific Northwest in 2021, they had analyzed nearly a dozen extreme events and their technique was well honed. It took Otto and her team exactly nine days to say that the heat wave that killed more than a thousand people and a billion sea creatures would have been "virtually impossible" without climate change.

Such bold statements, which also happened to be true statements that stood up under intense scrutiny from other researchers, made Otto a rock star in the world of science. She and van Oldenborgh were named to the 2021 *Time* 100, the magazine's annual list of the hundred most influential people of the year. *Nature,* the prestigious science journal, chose Otto as one of the ten people who made significant contributions to science in 2021.

"Attribution studies are really essential in terms of understanding human impacts of climate change," said Emily Boyd, a social scientist at Lund University in Sweden who studies climate adaptation and governance. "The science is shifting our mind-sets — it allows us to think about the relation between climate and vulnerability in a completely new way." More specifically, extreme event attribution is a tool that can profoundly reshape public conversation about the climate crisis. Instead of framing the crisis as a future event, something that will impact our children and grandchildren and generations to come, as it often is, Otto's work is proof that it is happening *now,* in real time. Among other things, real-time attribution will likely turn out to be a vital tool in court, opening the door to legal remedies for Allen's original question of legal liability: *Who is responsible for trashing the climate, and how can they be held responsible?*

As revolutionary as all this is, looking back at heat waves is easier than looking forward. Or to put it another way, just because it's now possible to say with certainty that human-made climate change has made heat waves more frequent, more intense, and more deadly, it doesn't help much to answer the next questions: How hot is it going to get? And where will the next heat wave hit?

Nobody expected a 70-degree jump in temperature during a heat wave in Antarctica in 2021. And yet it happened. Nobody expected 121 degrees in British Columbia. And yet it happened. Nobody expected 104 degrees in London. And yet it happened. As of 2022, the current record high in Phoenix is 122 degrees. Could it hit 135 degrees? How about 140 degrees? If not in Phoenix, how about in Pakistan? What are the limits? I talked to a number of scientists, who all pointed out the heat waves can be amplified by local conditions, from how dry the soil is to how much pollution there is in the air (the particulates that make up smog, paradoxically, act as tiny mirrors, reflecting away sunlight and keeping places cool) to hotspots in the ocean. But every scientist I talked to agreed on one thing: the more fossil fuel we burn, the higher the extremes can get.

But perhaps the more urgent question is: How hot can it get *right now*? Or, to put it another way, with today's level of warming, are there any brakes on the atmospheric system that prevent a heat wave beyond any we've experienced or imagined?

To answer this question, I have to take you on a brief detour into the world of atmospheric dynamics, which is the next frontier of heat wave science. Atmospheric dynamics is a fancy way of talking about how the air above us moves around the planet, creating our weather. One way to think about atmospheric circulation is as a giant heat transport system, one that is constantly circulating warm air from the tropics up to the poles, and bringing the cooler air from the poles down to the tropics. The main engine of this heat transport system is called the jet stream, which blows west to east in the upper atmosphere. TV weathercasters love to talk about the jet stream, often illustrating it with red (for warm) and blue (for cool) arrows circulating over images of the Earth.

Heat waves are created most often by changes in the jet stream. The pressure waves that guide the jet stream, known as Rossby waves, are shaped by the temperature differences between the poles and the tropics. Think of them as guardrails for the jet stream, boundaries that keep it on track. The Arctic, however, is warming four times faster than the rest of the planet, due in part to the melting ice that allows the oceans and the open land to absorb more heat. (Ice is a very good reflector, bouncing away sunlight and cooling itself. But eventually, the warming melts enough ice to expose heat-absorbing land or water, which in turn accelerates the ice melt even further.) As the Arctic warms, it's changing the temperature gradient between the poles and the tropics. That in turn weakens the Rossby waves, allowing the jet stream to meander and get twisty. Sometimes those twists trap hot air over a region, not allowing it to escape. The trapped air gets hotter and hotter, as a result of both the warm land below and the increasing high pressure that keeps out clouds and amplifies the sunlight. And that, to oversimplify greatly, is how you get a heat wave.

Among the big questions scientists are grappling with now: As the world continues to warm, how weird and wobbly will the jet stream get?

One example: A few weeks before the heat wave hit London in 2022, researchers published a study that raised the alarm about the risk of heat waves in Europe. The authors found that Europe is a heat wave hotspot, where heat waves are warming three to four times faster here than on average in the midlatitudes. Why? The jet stream is splitting apart over Europe, leading to more high-pressure zones. Double jet streams, or split flow as meteorologists like to call it, happen naturally. But the authors found that they are increasing and may explain a large portion of the faster acceleration in heat intensity over Europe. A meandering jet stream was also implicated in the 2021 Pacific Northwest heat wave that I described in the prologue.

"Everybody is really worried about the implications of these recent extreme events," van Oldenborgh said shortly before his death from cancer in 2021. "This is something nobody saw coming, nobody thought possible. We feel we do not understand heat waves as well as we thought we did."

So the best answer the scientists I talked to could give me to the question of how hot can it get right now is . . . well, they don't know. But it is a subject of active investigation. "I don't think anyone believes you could have a seventy-degree jump in New York City," Otto explained. "But five degrees above the all-time high? Ten degrees? That certainly seems within the realm of possibility."

The more extreme heat waves become, the more deadly they will be to people who are unprepared and vulnerable. And the more extreme they become, the more the question that Myles Allen asked himself twenty years ago will resonate: Who is responsible for this? At some point in the not-so-distant future, the question of who burned the fossil fuel that caused the heat wave that killed Jane Doe will become the climate version of who pulled the trigger of the gun that killed Jane Doe. And when the science and the law advance enough that Otto and her colleagues convince a judge that they can answer that question, a new age of accountability will begin. If that sounds like no big deal, consider this: it means that a company like ExxonMobil, which, by some measures, is responsible for about 3 percent of historic global CO_2 emissions, could be sued for 3 percent of the deaths or property destruction and economic losses from every climate-driven flood and heat wave — past, present, and future. To say that there are hundreds of billions of dollars at stake doesn't begin to describe it. This is one reason why the discussion about "loss and damages" is so fraught in international climate talks, with the leaders of rich, industrialized nations in the Global North pushing hard to keep the topic out of the negotiations while politicians and activists in the climate-ravaged countries of the Global South say, very bluntly, "We're the ones who are suffering. We're the ones who are dying. You owe us, and you need to pay."

I asked Otto if she can imagine a day in the near future when a company like ExxonMobil is held liable in a court of law for the deaths in an extreme heat wave.

"Yes, I can," she replied without hesitation. "Not only can I imagine it. I believe it will happen sooner than you think."

6. Magic Valley

By July 2022, rancher Mickey Edwards in Lampasas County, Texas, had seen enough. The grass on much of his land, he said, was "crunchy, dry, and dusty." It was useless for his cattle, who were growing thin and bony. For a while, he fed them bales of hay from last year's harvest — until a stock pond went dry, and the cattle had nothing to drink. He ended up selling off about forty animals, or more than 15 percent of his herd.

John Paul Dineen in Ellis County near Dallas said his seven hundred acres of corn were "pathetic and scraggly." He'd walk out in the field and peel back an ear and it was dry and hard and only half covered with kernels. He knew it was going to be a bad year. He began harvesting his corn in mid-July, joining many corn growers in the state who started a couple of weeks earlier than usual because the extremely high temperatures caused plants to mature ahead of schedule.

But there wasn't much corn, however pathetic and scraggly it may have been, for Dineen to harvest. The stalks were only five feet tall at most, compared with six to seven feet normally. "Yields are fifty percent of an average year," he said. "We might get fifty bushels per acre," which was far below the usual 100 to 110 bushels.

Dineen was not the only corn grower who was suffering. By midsummer, according to the US Department of Agriculture, 42 percent of Texas corn acreage was in poor or very poor condition. Only 3 percent was in excellent condition. David Gibson, executive director of Texas Corn Producers, a trade group, said the heat and the drought had been devastating for many farmers even though the market price for corn was high. "When you've got no corn to sell, a good price doesn't keep you in business," Gibson said.

It wasn't just Texas. In 2022, extreme heat hurt crop harvests all over the world. The corn harvest in France was the lowest in three decades. Throughout the European Union, soybean and sunflower yields were projected to be 10 percent lower due to extreme heat. In India, the wheat crop was significantly below projections, leading the government to put a ban on wheat exports, alarming grain traders and food security analysts. Devinder Sharma, India's leading agriculture expert, defended the government's export ban, arguing that the country must ensure it had enough food for its 1.4 billion people before it sold wheat abroad. "Look at what the heat wave did to our crops this time," Sharma told CBS News. "Who will be responsible if monsoon rains wreak havoc too, or if some other climate factors hit our production next year?"

Without enough food, there is only hunger, chaos, and violence. Malevolent rulers and dictators throughout history have known this and exploited it for their own ends. Russian president Vladimir Putin not only understands the political power of food, he also weaponizes it. With the Russian invasion of Ukraine in 2022, he deliberately disrupted the country's wheat supply, triggering a global food crisis. Until the invasion, Ukraine was one of the top wheat exporters in the world. By blockading Ukrainian ports, blowing up rail lines, stealing grain, and killing farmers, Putin effectively took about 20 million tons of wheat off the market. Global wheat production is about 850 million tons, so that was hardly enough to cause a worldwide famine. But it was enough to cause the price of wheat to jump more than 60 percent. In the US, where the average American spends less than 10 percent of their income on food, rising food prices put the squeeze on many working families and became a hot political issue in the 2022 midterm elections. But for people in the developing world, many of whom spend 40 percent or more of their income on food, a big jump in the price of wheat was the difference between eating and going hungry. In 2022, high food prices sparked riots in Sri Lanka and pushed an additional twenty-three million people toward starvation in sub-Saharan Africa. "By attacking Ukraine, the breadbasket of the world, Putin is attacking the world's poor, spiking global hunger when people are already on the brink of famine," Samantha Power, the administrator of the United States Agency for International Development (USAID), told reporters.

This war-driven food crisis, however, was in some sense artificial, given that it was not driven by any actual shortage of food in the world. Even with the Ukrainian wheat off the market, there was still plenty of grain to go around. The issue was how much it cost and how it was distributed. And Putin was not the only one who was exploiting this situation. Commodity traders make money off wild price swings, shippers make money off people desperate for grain, fertilizer manufacturers make money off farmers desperate to maximize their yields, and protofascist politicians are happy to exploit rising food prices as evidence of the failure of democracy.

Behind the immediate wartime food panic, a much bigger and more worrisome crisis loomed. Since 2019, the number of people facing acute food insecurity has soared — from 135 million to 345 million. In 2022, 50 million people in 45 countries were teetering on the edge of famine. Whatever the virtues of our modern food system may be, eliminating global hunger is clearly not one of them. More than 30 percent of the food we grow is wasted, left to rot in warehouses or tossed out by finicky consumers who decided they didn't like the sauce on their pasta. In the US, 30 million acres of prime agricultural land is devoted to growing corn and soybeans to make fuel (mostly ethanol) for gas-guzzling cars and trucks. Aquifers are being pumped dry, often to grow water-hungry crops like rice and almonds and alfalfa in places where there is little or no surface water. In northern India, one of the prime food-growing regions of the country, groundwater is being pumped out so fast that the water table is falling by about three feet a year.

In the coming years, the challenge of feeding the world will only get more complex. For one thing, the world population is projected to grow from 8 billion today to nearly 10 billion by 2050. To meet the expected demand for food by midcentury alone, global agricultural output will have to rise by more than 50 percent. How will that work? The World Resources Institute estimated that meeting that level of agricultural output would require clearing at least 1.5 billion acres of forests, savannas, and wetlands for new farmland, an area nearly twice the size of India. That's obviously not going to happen, but it does give you a sense of the scale of the problem.

Meanwhile, food productivity is already in decline due to human-made climate change. One recent Cornell University–led study found that global crop production today is 21 percent lower than it would have been without climate change. The losses were higher for warm regions — such as Africa, Latin American and the Caribbean — than for cooler regions such as North America and Europe. But as long as the heat keeps rising, the overall decline in crop productivity is likely to continue. For every degree Celsius of increase in global mean temperature, yields are expected to decrease by 7 percent for corn, 6 percent for wheat, and 3 percent for rice.

It is true that farmers have had to deal with wild weather for as long as humans have been putting seeds in the ground. But this is different. This is not about freaky hailstorms and random cold snaps. As Donald Ort, a professor of plant biology at the University of Illinois Urbana-Champaign, explained to me, "The largest single global change that threatens food security is high temperature."

The risk to food security from extreme heat begins with basic physics and biology. Like humans, plants live in their own Goldilocks Zone. They respond to temperature just like humans do. Except they can't crank up the air-conditioning or cruise over to the beach if it gets too hot. They are stuck where they are. Yes, plants can move to more suitable climates over time, especially those that regenerate by seeds that can be transported in the wind or dispersed by birds or enterprising humans. Given enough time, entire forests can migrate to cooler climates. But individual plants, once they take root, are stuck where they are. If it gets too hot, they are in trouble. Extreme heat waves are particularly dangerous because they come out of nowhere and the plants have little time to adapt.

Heat increases the metabolism of plants, just like it does in humans. It raises their heart rate, in effect. And that speeds up everything, including the need for water. Plants are about 97 percent water (humans, in contrast, are 60 percent water). Water is key for all their basic functions, including photosynthesis. Some plants are more efficient than others in dealing with limited water. The prickly pear cactus, which thrives in the deserts of northern Mexico and the southwestern US, stores water in its stems and protects itself from thirsty predators with long, needlelike spines. Pistachios in California's Central Valley can thrive with far less water than almonds. But however well adapted a plant may be, the water/heat relationship is absolute: the hotter it gets, the more water it needs. "Plants are water-pumping machines," one biologist told me.

When it gets hot, plants do more or less what humans do — they sweat (in plants, it's called transpiration). Instead of sweat glands, plants have tiny openings in the underside of their leaves that release water vapor, similar to skin pores. Most plants transpire their weight in water every day (if humans sweated that much, we'd have to drink twenty gallons of water a day). For plants, even small changes in temperature mean big changes in sweat. "To get a sense of how important temperature is, if you go from 25 to 35 Celsius [77 to 95 degrees Fahrenheit] you more than double the amount of water needed to maintain a given level of growth," says David Lobell, an agricultural ecologist at Stanford University. Corn is a particularly big sweater. In the summertime, a single acre of corn in Iowa can sweat four thousand gallons a day — enough to fill a typical residential swimming pool in less than a week. This need for water is why, for crops like corn and soybeans, dry heat is much more damaging than wet heat — dry heat not only sucks the moisture out of the plants, but it is usually accompanied by lower rainfall, which dries out the soil that allows the crops to thrive.

Heat impacts plants in other ways. It changes the timing of blooms, which can put them out of sync with pollinators. Rising heat also makes plants more vulnerable to diseases like aflatoxin, a fungus that grows on corn and can kill you with one bite. In a hotter world, rice — an important source of nutrition for hundreds of millions of people around the world — sucks more arsenic out of the soil, creating arsenic-laced kernels (arsenic-infused rice won't kill you, but chronic exposure has been linked to breast and bladder cancer, as well as neurological issues in young children). Heat also amps up the life cycle of pests that attack plants. Instead of maturing in twenty-eight days, caterpillars might mature in, say, twenty-one days. More rapid maturity means more generations of pests in a season, multiplying the damage they can do.

For years, the fossil fuel industry argued that burning coal, oil, and gas was good for plants. After all, fossil fuel combustion releases CO_2; as everyone knows, plants breathe in CO_2 the way humans need oxygen. In greenhouses, a little dose of CO_2 can accelerate plant growth. And it's true that as CO_2 levels in the atmosphere increase, the Earth is getting greener.

But it's more complicated than that. Like humans, plants also acclimatize, so the effect diminishes with time. More CO_2 also means more heat, and the effects of heat quickly overwhelm the benefits of higher CO_2. It also makes some plants less nutritious. Rice grown in high-CO_2 conditions has lower amounts of protein, iron, zinc, and B vitamins.

Plants, like people, are not equally vulnerable to rising heat. Wheat can move to higher elevations. Pearl millet, an important grain in hot, dry regions of India and Africa, might thrive in other hot, dry places, including the American Southwest. Irrigation systems can be improved to use less water. New varieties can be bred to be more heat- and drought-tolerant. But our world is changing fast. How and where we get our food is not.

In the nineteenth century, early white settlers considered the Rio Grande Valley, where the once-mighty Rio Grande River defines the border between the US and Mexico, to be a forest of thorns and cacti where tarantulas crawled into your boots and ocelots slinked by just beyond the campfire. Nomadic tribes of Native Americans had thrived in the valley for centuries, and both Spanish and Mexican soldiers had long been marching through the region. But among the waves of settlers from the east coast and the Midwest who came looking for farmland, few people thought the region would be good for much beyond cattle and fighting.

Then around the turn of the twentieth

century, somebody decided to dig ditches in the banks of the Rio Grande to let water irrigate the land. Because it was hot, they planted sugarcane, which grew well in the tropics. It flourished. And landowners in the valley realized that with a little water, this was an agricultural paradise. It was warm, which meant no big freezes and a long growing season, with plenty of sunshine and rich river-bottom soil so fertile it gave birth to farmers' most vivid agricultural fantasies. With water, they discovered, heat could be tamed.

Soon there were canals everywhere and grapefruit trees and tomatoes and corn and cotton and lettuce. Whatever you planted, it grew. Some civic booster came up with a snappy nickname for the Eden-like region: the Magic Valley.

And for the last hundred years or so, it has been magical. The Rio Grande Valley — which, in fact, isn't much of a valley at all, just a big flat expanse of river-bottom land created over centuries by the ever-changing Rio Grande — is one of the most agriculturally productive regions in America, growing everything from watermelons for Fourth of July picnics to sorghum for animal feed to papaya for salads and smoothies.

The original driver of the Magic Valley's

magic is an old irrigation pump in Hidalgo, a small town that is right on the border. The pump house itself was closed when I visited but I wandered on the grounds and saw the irrigation channel that had been dug more than a century ago. A historical marker explained that the pump house had been built in 1909 and powered by steam. It pulled water out of the Rio Grande and pushed it into a channel, which spread into small channels and ditches like veins all over the valley, eventually irrigating about forty thousand acres of land. The old pump ran until 1983, when it was replaced by a new all-electric pump downstream.

I hoped to see the Rio Grande from the old pump house, but there was no Rio Grande to be seen. Just a chain link fence, a border guard asleep in his patrol car, and, a few yards behind him, a towering steel wall. I learned later that the Rio Grande had shifted its course over the years, and it was now a mile or so farther west, out of sight from the old pump house.

Visible or not, the fate of this region is inseparable from the fate of the Rio Grande, one of the great and storied rivers of the West. It flows out of the mountains of Colorado and makes its way across New Mexico, then down to El Paso. It defines the US/

Mexico border all the way down to Brownsville, where it empties out into the Gulf of Mexico. In some parts, especially just below El Paso, the river vanishes entirely before being replenished by the Rio Conchos River, which flows out of the mountains of northern Mexico. By the time the Rio Grande gets down to the Rio Grande Valley, it's about thirty feet wide and nearly exhausted from its long journey.

When I visited the valley in 2022, the Southwest was suffering from the worst drought in twelve hundred years. According to the US Drought Monitor, exactly 99.8 percent of the region was classified as in "extreme drought." Decreasing snowpack in the Rockies has diminished the river's headwaters. Golf courses and condos along its banks, as well as farmers who grow water-intensive crops like pecans, leech its flow. The water that is left is increasingly salty and tainted by nitrogen runoff and other pollutants. As the heat rises, good water is needed more than ever. But there simply isn't enough to go around anymore. "When you cut the water off, everything changes. Not only nature, but you change people's way of living, the economy, the biodiversity," says Estela Padilla, seventy-seven, a retired state worker who has lived along the river her entire life. "It is so colossal. It's like a crime scene."

Alexis Racelis lives in a 1960s suburban-style house with a big mesquite tree and a shaggy vegetable garden in the front yard. When I arrived at 7 a.m., he was up and loading gear into his black Nissan Pathfinder. He's forty-six years old; he was wearing a khaki shirt and well-worn work boots, and if you didn't know who he was you would guess he was a farmer. He has a salt-and-pepper beard and broad shoulders and the genial, no-nonsense manner of a serious scientist. His parents were both born in the Philippines and immigrated to the US in 1976. Racelis grew up in San Diego and was close with his grandfather, who had been a nurse in the Philippines during World War II. "I used to follow my grandfather around in the garden," Racelis says. "He had a deep perspective of life and survival that seemed to emanate from everything he did." After getting his PhD at the University of California, Santa Cruz, Racelis went to work for the US Department of Agriculture, focusing on invasive species. Now he's an associate professor of agroecology at the University of Texas Rio Grande Valley. He still tracks invasive species, but he spends most of his time working with farmers to find ways to grow food with less water and less dependence on Big Ag for seeds and fertilizer. He's also a leader of a growing local food movement and founded a five-acre farm and community garden in the city of Edinburg called the Hub of Prosperity.

"We have a long growing season here, which has always been a big advantage," he told me as we drove out of town in his Pathfinder. "But in the summer, it's so hot that it's basically our fallow season." Nothing can fruit when it gets above 95 degrees, which it does from May to September. And the number of hot days is increasing. Racelis talked about the challenge of getting the crops in and out of the ground before it gets too hot in the spring. "If this heat keeps up," he said, heading out beyond the strip malls into agricultural land, "broccoli and cauliflower is going to start bolting, which is a big problem."

Racelis drove forty-five minutes across fields of onions and watermelons, as well as giant squares of land waiting to be planted with sorghum and corn, before we arrived at a seven-hundred-acre aloe farm. It was a beautiful place, with a row of palm trees on the main road and a big farmhouse nestled in some trees on a small hill. The temperature was a balmy 92 degrees. Not bad — except that it was February.

Out in the field, we met Andy Cruz, a forty-something man in rough, muddy boots who was in charge of growing things on these seven hundred acres. He had been a farmer in the valley all his life, he told me, and had grown everything from corn to cucumbers. Now he was in charge of seven hundred acres of aloe. Aloe is not primarily a food crop (although people do use it to cook, and for flavor in yogurts and desserts). It's grown mostly for the aloe vera gel that is found in the leaves of the plants, which has been used for thousands of years for skin care and other health benefits. It's probably in the soap you use to wash your hands and the sunscreen you wear when you're at the beach.

By all accounts, aloe should be the perfect crop for the valley. To begin with, it's a succulent — that is, a plant with thick, fleshy tissue that's well adapted to water storage. It evolved in the heat of Africa, and, like many succulents that emerged in hot places, it has developed a unique capability to in effect hold its breath during the heat of the day and breathe only at night when it's cooler. During the day, the aloe's stomata — the small mouthlike structures on the underside of plant leaves and stems — close up tight, minimizing the water loss that would otherwise occur while the plant is inhaling CO_2. Then at night, when the temperatures are cooler, the stomata open and the plant breathes.

Aloe has another, even more remarkable heat adaptation: if it gets too hot and dry for too long, the plant can put itself into what amounts to a temporary hibernation, slowing down its metabolism to the point that its water and CO_2 needs are minimal. Then when it rains, or sufficiently cools down, the plant wakes up and comes back to life.

Despite these wonders, the seven hundred acres of aloe have been a world of trouble for Cruz. For one thing, a plant that's well adapted to heat is not necessarily well adapted to the *lack* of heat. A cold snap in the valley in the winter of 2020 killed half the plants on the seven hundred acres. "It was bad," Cruz told me. "We were out here for two days and nights with burners trying to keep things warm. This climate change thing is making the weather like a Ping-Pong ball — you never know where it is going to bounce."

As he told me about this, we were walking among rows of aloe. I started to explain how some scientists hypothesize that the freeze that hit Texas in 2021 — which I froze through myself — was one of the consequences of a warming Arctic. I said, "The Arctic is warming four times as fast as the rest of the planet, and as a result, it is pushing the jet stream farther south, allowing the cold arctic air to move down to Texas." I wanted to add that at this very moment, the temperature in the Arctic was again fifty degrees above normal, which meant Texas could freeze again. But I stopped. Cruz was staring down at his well-worn, muddy boots and I suddenly heard myself talking like some kind of wise-ass city guy. Cruz clearly knew a million times more about the relationship between plants, heat, and life than I did.

"Up until five years ago, things were fairly predictable," Cruz told me. "But now, you never know what's coming. It's different. Something's changed."

Of all the commercial food crops, corn may be the most vulnerable to heat. It does have some advantages, however. Most plants use what's known as C3 photosynthesis to convert sunlight into food (it's so named because the carbon compounds produced contain three carbon atoms). But C3 photosynthesis is problematic because about 20 percent of the time, these plants make a mistake and instead of a carbon molecule they grab an oxygen molecule, which is useless to them.

Corn is a C4 plant (others include sorghum and sugarcane), which uses a different process that avoids the mistake of confusing oxygen with carbon, which makes the plant's metabolism more efficient. Also, similar to aloe and other succulents, it closes up its stomata on hot days, allowing it to conserve water and better tolerate heat (corn moderates its breathing depending on temperature, while aloe breathes only at night, regardless of temperature).

Like aloe and other succulents, corn evolved in a warm place. Its wild ancestor, a grass called teosinte, thrived for ten thousand years in the Balsas River Valley in south-central Mexico, where the temperature is a steady 80 degrees. That means that deep in its ancestral gene pool, it has more tools to handle heat than many plants.

But 80 degrees is very different than, say, 102 degrees. As the world heats up, corn is nearing the limits of its adaptive (or "permissive") temperature range. To put it another way, it's already growing in hot places and now those places are getting hotter. Add a modest heat wave to those already hot places and corn has trouble coping. Add an extreme heat wave and it may not recover at all. Corn is in particular trouble if that heat wave hits during the reproductive cycle. "Heat disrupts the development of the pollen tube, which is deployed from the pollen to deliver the sperm into the female plant's ovule," plant scientist Donald Ort told me. "So the plant never fertilizes, there is no ear."

Corn is also vulnerable because, at least here in the US, big seed companies like Monsanto and BSF have had their way with the corn genome for years. Growing it requires huge amounts of nitrogen fertilizers, which end up polluting rivers and lakes, causing huge algae blooms. In the right conditions, commercially bred corn is a superstar. But if heat spikes and rains don't come at the right time, it's vulnerable. In the highly engineered varieties of corn that most farmers grow, much of its rich genetic diversity has been bred out and what is left is an army of high-yielding inbred hermaphrodites, exquisitely suited for a narrow range of conditions that exist in the Corn Belt. Or that existed there before climate change started making a mess of the weather.

So why not just plant corn in cooler places? It's not so simple. "If you plant corn in the Central Valley in California and you give it unlimited water, it does amazingly well," Jeffrey Ross-Ibarra, a corn geneticist at the University of California, Davis, told me. "It's just not economically feasible. The profit margins are such that you'd rather be growing grapes or almonds or something in California if you're going to spend a lot of money on water. So I don't think it's as simple as saying temperatures are going to push everything north and we'll be fine, because it is a combination of soils and water and government regulations and agronomic practices and even contract preferences. One of the issues in Mexico, for example, is there's corn varieties for specific foods. If you hand a farmer a particularly well-adapted, heat-tolerant corn variety that is terrible for pozole, they're not going to grow it if what they want is corn for pozole."

The vulnerability of corn matters because it's the industrial food stock of American life. Processed foods, from breakfast cereals to ice cream, are saturated in corn syrup. Corn is also a prime feedstock for animals, which have themselves been engineered to ingest and digest huge amounts of corn and transform it into animal protein. In this sense, a McDonald's hamburger is better thought of as a McDonald's corn burger. Corn fuels your trip to McDonald's, too. More than half the corn grown in Iowa actually ends up as ethanol, which is mixed with gasoline and is an essential ingredient in fuel.

If corn productivity declines, there will be more pressure to clear land to grow more, which is bad news for places like the Amazon rain forest. It will also raise prices for many food staples, especially meat. How exactly that will play out will be different in different places at different times. But as the Russian invasion of Ukraine demonstrated, higher food prices are inextricably linked with political instability, chaos, and war. Sharply rising prices were a major driver of the French Revolution. Food protests not only kick-started the 1917 Russian Revolution from which the Soviet Union was born but also, ironically, contributed to the USSR's demise. The Arab Spring uprising, which began in 2010 and jolted the political stability of the Middle East, was triggered in part as a protest against rising food prices.

Racelis and I drove out to a twelve-thousand-acre ranch on the western edge of Hidalgo County. He wanted to check in on an experiment he was running that measured the effectiveness of cover crops at keeping moisture in the ground. The temperature had dropped forty degrees overnight and there was a slight drizzle, which, given the epic drought the region was facing, was a good sign. Racelis talked about how farmers were waiting for one heavy rain before putting crops like cotton and sorghum in the ground. "Growers have been informed that there is enough water for them to have one good watering this year," Racelis explained. "Usually they get two or three. So if they can, they want to wait to plant until we get a good rain, so they can save their irrigation water for later in the spring when they really need it."

As we drove across the misty fields, I asked him about his discussions on climate change with growers in the valley. He told me it is not something they are eager to talk about — "it usually only happens after several shots of tequila," he told me.

"That said, growers here are very resourceful," Racelis explained. He talked about how farmers practice what amounts to a just-in-time adaptation strategy, shifting to different crops and planting times as conditions change. But at a certain point, I asked, when the heat rises and the water runs out, what will happen to agriculture in the valley? "Things like okra will always grow by the river, as long as the river is flowing," he responded. "But how much okra does the world need?"

After about an hour, we turned down a narrow farm road. In the distance, I saw three or four people out in the field, a few trucks pulled over on the side. We parked beside them, then walked into the field. Racelis introduced me to a few of his students, then to some farmers he was working with. One of them was Avan Guerra, a weathered-looking man who wore his jeans tucked into his boots. He is a former parole officer who got into farming and now has three hundred acres. Guerra owns his own tractor and other equipment. He grew up here, and remembers the heat when he was a kid, but it didn't seem that bad back then. "Nobody used to need air-conditioning to live around here," he said. The erratic weather is not just making it tougher for his crops. It's also making it tougher for him to decide every year what to put in the ground. "I hope to get another decade out of this, and then I'm heading to Vegas," Guerra told me, smiling.

Scientists have tools now that allow them to cut and paste DNA almost as easily as I can cut and paste the words on this page. CRISPR, as the technology is called

(Clustered Regularly Interspaced Short Palindromic Repeat), is revolutionizing agriculture and helping create crops for the future. Sounds promising, right? Just insert a cactus gene into the corn genome and *voilà!* you have supercorn, capable of withstanding the heat of a thousand suns.

But it doesn't work that way. Heat is not a trait, like blue eyes. "Understanding how plants react to heat on a genetic level is like trying to understand cancer," Meng Chen, a researcher at UC Irvine who is working on heat tolerance in plants, told me. "You mess with one thing, you mess with everything."

A good example: how corn reacts to light. "One of the things that took a long time for corn to adapt to as it moved north from Mexico was a change in sunlight," Jeffrey Ross-Ibarra explained to me. "The plant is used to twelve-hour days, twelve-hour nights. If you try and grow it in California's Central Valley, the plants freak out because the day length is inappropriate. And instead of starting to flower when they should, they just keep growing and you can end up with a corn plant that's twenty-five feet tall."

Other researchers are skeptical about CRISPR-manipulated plants for a different reason: even if they can work and solve heat-related problems, the seeds of the modified crops will be locked up by big seed companies, furthering the corporate control of farmers and our food supply. It is not going to help people in, say, Madagascar, where they are starving, or in Colorado, where people just want to grow some food in a community garden.

Other researchers are exploring the genetic diversity acquired in millions of years of evolution. Because corn evolved in a hot place, researchers suspect that there is surely a sequence of genes that make some varieties of corn more resilient to heat than others — but how do you find them? "We can find genes for simple traits, but for anything complicated like yield or heat tolerance, it's just not going to happen," Seth Murray, a plant geneticist at Texas A&M University, told me. "There's so many different genes in the genome, and they're all interacting. I calculated that we would have to grow more corn plants than there are stars in the sky and measure all of them to figure out the function of all the genes in the genome." Instead, Murray searches for traits like heat tolerance by planting thousands of varieties and using drones to see which varieties grow best. It's a way of exploring the genetic diversity that is buried in various strains of corn and other crops without having to map the DNA itself.

Other researchers are looking for ways to replace corn with an entirely different crop, such as Kernza. Unlike corn, which has to be replanted every year, Kernza grain is harvested from a wheatgrass perennial that continues to grow for a decade or two after it is planted. Right now, Kernza is often used as a substitute for wheat in bread and beer production, besides being used as a whole grain like barley or rice. Kernza, as well as other perennials like legumes and oilseeds, are more drought- and heat-resilient because of their root structures. Unlike the shallow roots of corn and wheat, Kernza has a large root system that can reach ten feet into the ground, helping it suck up water even in harsh conditions. And Kernza is hardly unique. "There are so many other potential new crop species out there that could be far more drought-intensive or heat-tolerant than even any of the ones we're working on right now," says Tim Crews, chief scientist at the Land Institute, a nonprofit agricultural research group in Salina, Kansas, where Kernza was bred (and trademarked) from wild wheatgrass.

Another way to grow food on a hot planet is to simply move indoors. Several years ago,

I met a guy named Jonathan Webb at a tech conference in Idaho. He had a dream about building a giant indoor greenhouse in Kentucky to both create jobs and grow food far more efficiently, and with a far smaller carbon footprint, than the old-fashioned way of sticking it in the ground, hiring laborers to harvest it, then loading it into eighteen-wheelers and trucking it across the country to a supermarket near you. I thought it was a noble but crazy dream, given that Webb's background was in solar, not agriculture. He knew as much about growing tomatoes on a commercial scale as I did.

Flash forward five years: Webb and I are walking through a sixty-acre greenhouse near Morehead, Kentucky. His dream had become real: AppHarvest, the company he started shortly after we talked, had gone public and had a market cap of $500 million. "The old way is broken," Webb told me. "This is the future of food." Inside, it felt like a jungle, but a well-organized one. Thousands of tomato plants grew on scaffoldings, their roots in pods of water. Each one is watched by computers that precisely monitor the needs of each individual plant. The greenhouse uses 100 percent recycled rainwater. No chemicals, no pesticides, no agricultural runoff. LED lights overhead provide the sunlight on cloudy days. Temperature is precisely controlled. And this whole system is cloneable and will only become more efficient over time. When I visited, Webb was overseeing the construction of two other greenhouses in Kentucky, one to grow berries, another for leafy greens. Promising as all this may be, you have to be a pretty hard-core technofuturist to imagine growing enough corn and wheat under glass to feed millions of hungry people.

Then there's protein. Besides being major sources of greenhouse gases themselves, cows and chickens and pigs are all extremely vulnerable to heat. During the summer of 2022, thousands of cows died from heat stress in Kansas feedlots and had to be buried in landfills and hastily dug pits. In Texas, increasing heat is encouraging the return of Texas cattle fever, a deadly tick-borne disease that can wipe out herds and that ranchers spent decades trying to eradicate. As temperatures rise, keeping animals cool will become more difficult and expensive. Air-conditioned cows, anyone? The transportation of animals in hot weather is also deadly and dangerous. In 2019, twenty-four hundred sheep were boiled alive when an ocean transport ship was delayed near Kuwait in 100-degree heat. Other protein alternatives are emerging fast:

cell-based meats grown in labs; plant-based proteins that are engineered to mimic meat, like Impossible Burgers; fungi-based proteins created by microbes discovered in the acidic hot springs at Yellowstone National Park; cricket farms capable of producing fourteen thousand tons of crickets a year, which can be ground up and made into a protein-rich flour, or seasoned and fried like shrimp (I've eaten *chapulines* several times in Mexico, where they have been a traditional food for centuries — they're delicious). Replacing feedlots with cricket farms and lab-grown proteins has lots of co-benefits too. They would not only reduce the animal suffering associated with slaughterhouses, they would also free up vast amounts of land for wildlife and forests.

But we are not going to feed a hot planet with ten billion people with cricket farms and lab-fermented protein. At least not for a while. We need to maintain a crop-friendly planet. But as the heat rises, the dangers of breaking our global food system rise too. "People will shift crops around, try new varieties," Racelis told me as we drove around the valley one day. "But in the end, there is no getting around the laws of physics and biology. When it gets too hot, things die. That's just how it works."

7. The Blob

The Blob went unnoticed at first. In the summer of 2013, a high-pressure ridge settled over a Texas-size area in the northern Pacific, pushing the sky down over the ocean like an invisible hand. The winds died down, and the water became weirdly calm. Without waves and wind to break up the surface and dissipate heat, warmth from the sun accumulated in the water, eventually raising the temperature by five degrees — a huge spike for the ocean.

When scientists noticed this temperature anomaly in the satellite data, they had never seen anything like it. Everyone knew about heat waves on land, but in the ocean? "As the Earth heats up, the ocean is changing in very dramatic ways," Jane Lubchenco, a marine ecologist and former head of the National Oceanic and Atmospheric Administration, told me. "It is less predictable, and we are seeing more surprises.

The heat waves are one of those surprises."

Nick Bond, a climatologist at the University of Washington, nicknamed the Pacific heat wave the Blob, after a campy 1958 sci-fi movie about a gelatinous monster that arrives on Earth in a meteor and eats up a small town. But this Blob would turn out to be far more deadly than anything Hollywood had imagined.

The hot water killed the phytoplankton — a form of microscopic algae — that live in the top few hundred feet of the ocean. The tiny organisms that feast on them starved, including krill, the small shrimplike creatures that swarm the ocean by the billions and are the preferred food for whales, salmon, seabirds, and many other creatures. The population of herring and sardines, an important food source for many larger fish and marine mammals, also declined. By killing phytoplankton, the Blob disrupted the entire Pacific food chain.

Over the next two years, it drifted down the coast of Alaska to California, eventually becoming responsible for thousands of whale and sea lion strandings on beaches along the coast; the collapse of the Alaska cod fishery; the bankruptcy of fishermen and worker layoffs at fish-processing plants; the vanishing of great kelp forests on the Pacific coast; and the starvation and death of a million seabirds — the largest single mass mortality of seabirds ever recorded. Dead murres littered beaches like washed-up plastic bottles.

And its destruction was not limited to the ocean: the Blob changed the weather on the Pacific coast, pushing heat inland and altering rainfall patterns, contributing to the California drought. "It raised temperatures on the coast all the way from British Columbia down to Southern California," Daniel Swain, a climate scientist at UCLA, explained when I called him to ask about the trajectory of the Blob.

One question, still unresolved, is how much the Blob accelerated wildfires. The year 2018 marked the beginning of a series of historic blazes that burned millions of acres, including the Camp Fire in Northern California, which killed 85 people and displaced 50,000 others. Swain said the Blob increased nighttime temperatures in the western third of the state, where many of the wildfires flared. "Firefighters will tell you that's really important, because wildfires often die down at night, burning more slowly and behaving less erratically, becoming less dangerous to approach for human crews. While the Blob was off the coast, that didn't happen."

All in all, the Blob was a slow-rolling climate catastrophe. It's also compelling evidence of how tightly all life on Earth is linked to the ocean. Because we live on land, we often think of heat as a terrestrial event. But as temperatures rise, it's what happens in the ocean that may have the biggest impact on our future.

When water gets hot, every six-year-old knows, it boils. But long before that, it has other effects too. Heat causes water to expand (the faster molecules vibrate, the more space they need). It also changes how it moves: cold water sinks, warm water rises. If you think about the water in your bathtub, this might not seem like a big deal. But if you think about it on a planetary scale, it is a very big deal.

Water arrived on Earth from the cold depths of space with icy asteroids and comets, which bombarded the planet during the first few million years of its existence. It's been a watery world ever since. Today, 97 percent of the Earth's water is in the ocean, which covers more than 70 percent of the planet. The ocean was the petri dish for the evolution of life, and we carry that early history within us. The salt content of our blood plasma is similar to the salt content of seawater. “The bones we use to hear with were once gill bones of sharks,” explained Neil Shubin, professor of anatomy at the University of Chicago and author of *Your Inner Fish: A Journey into the 3.5 Billion-Year History of the Human Body*. “Our hands are modified fish fins, and the genes that build our basic body architecture are shared with worms and fish.”

Despite our intimate connection to the sea, for most of human history the ocean has been as strange to us as a distant planet, a realm of monsters and mayhem. Humans stuck close to the shore, mostly, and our ignorance about the ocean was profound. It still is. Scientists have only a vague understanding of how ocean currents are driven, or how ocean temperatures impact cloud formation, or what creatures thrive in the depths. Far more people have been to the moon, which is 240,000 miles above us, than have been to the deepest part of the ocean, which is 7 miles below us. Eighty percent of the ocean remains unmapped, unobserved, unexplored. Marine biologists still don’t know how sharks sleep or how an octopus learns to open a jar.

But scientists know enough to know that the ocean is undergoing profound changes. Largely because of overfishing, 90 percent of the large fish that were here in the 1950s are now gone. One metric ton of plastic enters the ocean every four seconds (at this rate, there will be more plastic than fish in the ocean by 2050). But the biggest problem is that the ocean is heating up fast. Since the 1960s, the rate at which the top mile or so of the ocean is heating up has doubled. In 2022, the ocean hit its warmest temperature on record for the fourth year in a row. By one measure, the amount of heat being added to the ocean is equivalent to every person on the planet running a hundred microwave ovens all day and night.

Until now, the ocean has been the hero of the climate crisis — about 90 percent of the additional heat we've trapped from burning fossil fuels has been absorbed by it. "Without the ocean, the atmosphere would be a lot hotter than it already is," Ken Caldeira, a senior climate scientist with Breakthrough Energy in California, told me. But the heat the ocean absorbed has not magically vanished — it's just stored in the depths and radiated out later. By absorbing and slowly releasing heat, the ocean reduces the volatility of our climate, cushioning the highs and lows as temperatures change from day to night, winter to summer. It also means the heat will continue to seep out for centuries to come, slowing any human efforts to cool the planet.

"The ocean is the main driver of our climate system," German climatologist Hans-Otto Pörtner told me. One of the central functions of the ocean, Pörtner said, is to redistribute heat from the tropics toward the poles via deep currents like the Gulf Stream system, which begins in the Southern Ocean near Antarctica and flows across the equator, up to the Arctic, and back again. "Even small changes in that system can have large impacts on things like the size and intensity of storms, rainfall patterns, sea-level rise," said Pörtner, "and of course the habitats of all the creatures that live in the ocean." A good example of how a warming ocean impacts the weather was the intense sequence of storms that battered California in early 2023, causing widespread flooding and mudslides throughout the state. The storms were driven by what scientists call atmospheric rivers, which carry moisture from the tropics northward. A hotter ocean — in particular, a hotter upper ocean — only intensifies these rivers in the sky. According to Kevin Trenberth, a climate scientist at the National Center for Atmospheric Research in Colorado, the California storms were "a direct consequence of the upper ocean heat content anomaly."

The ocean is also one of the main drivers of many regional economies. In Alaska, the seafood industry employs more than 62,000 workers, earning $2 billion in total annual income. Across the US, 1.8 million people work in commercial and recreational fisheries, contributing about $255 billion to the US's annual gross domestic product. No one thinks this blue economy is going to vanish tomorrow, but as fish and other species migrate to cool waters or die off from temperature changes, there can be big impacts on local fisheries — just ask the cod fishermen in Alaska, or shrimpers in the Gulf of Maine, who have been wiped out by rapidly warming waters in the Atlantic.

Pörtner is one of the lead authors of a 2019 report on the ocean and cryosphere by the UN's Intergovernmental Panel on Climate Change (IPCC). It was the IPCC's first report to focus specifically on the world's oceans and ice — it was a massive project, the work of 105 scientists over a three-year period. There is a lot of nuance in the report, but the basic message is clear: In the coming decades, the ocean will get hotter, more acidic, with less oxygen and less biodiversity. Seas will rise, flooding coastal cities. Ocean circulation patterns will shift, driving big and unpredictable changes in the weather, with scary implications for the global food supply. The report's summary was blunt: "Over the twenty-first century, the ocean is projected to transition to unprecedented conditions."

Monterey Bay is a crescent on the Northern California coast, a place haunted by the ghosts of John Steinbeck's *Cannery Row.* The old sardine canneries are now T-shirt shops and touristy restaurants. From the pier, you can watch sea otters playing in the surf and, if you're lucky, whales breaching just offshore. A deep canyon delivers cold, nutrient-rich waters into the bay, creating one of the most diverse ecosystems in the Pacific, including giant kelp beds that grow along the coast all the way up to Alaska. In good times, these kelp beds are teeming with life — otters, seals, sharks, rockfish, lingcod. "The kelp beds are the rain forests of the Pacific," Kyle Van Houtan, the former chief scientist at the Monterey Bay Aquarium, told me.*

* Unlike what grows in rain forests, however, the giant kelp in Monterey Bay are not plants. They are a type of brown algae, one of the oldest and simplest forms of life on Earth.

But like everything in the ocean, the kelp beds are changing fast. One Saturday morning, my seventeen-year-old daughter, Grace, and I pulled on our scuba gear and jumped in the chilly water near Monterey to have a look. I asked her to come along because I wanted her to see as many of the wonders of the world as she could while they were still around. She is also a great and fearless underwater companion. ("When we're diving, I always know when there's a shark around," I joke with friends. "Because Grace starts swimming fast — *toward* the shark.")

This dive was indeed wondrous. Instantly, we were on another planet. The giant kelp plants swayed in the push and pull of the tide. Sunlight filtered through the blades, giving everything a greenish hue. Blue rockfish and blacksmiths schooled around us. Grace pointed to an otter that darted over to see why we were trespassing in its neighborhood, then quickly disappeared. "That was incredible," she said as we hauled our dive tanks up onto the beach afterwards.

My second dive in another part of the bay, which I did alone with a dive guide, was entirely different. I was greeted with nothing but rock and water and hundreds of purple sea urchins, their thorny spikes like medieval armor. A voracious horde had invaded the once-magnificent kelp forest and devoured everything ("Purple urchins are the cockroaches of the sea," one scientist told me), leaving only some empty abalone shells, a rockfish poking around, and a few pathetic kelp stipes. And this spot is just one fragment of a bigger picture. As a result of the Blob, many of the kelp forests along the coast from California to Oregon have vanished, done in by warming and the army of purple sea urchins, which thrive in a hotter world.

"If a two-hundred-mile-long stretch of forest in the California mountains suddenly died, people would be shocked and outraged," Laura Rogers-Bennett, a marine scientist with the California Department of Fish and Wildlife, told me when I visited her at the Bodega Marine Lab a few days after our dive. "We're talking about the collapse of an entire ecosystem. But because it happened in the ocean, nobody notices."

Rogers-Bennett was one of the first scientists to understand the impact of marine heat waves like the Blob. In 2013, she was diving in Northern California when she saw a sea star that looked like it was melting. "When I touched it, its skin came off in my hand," she recalled. And it wasn't just one sea star, she discovered. This was the beginning of a mass die-off of twenty species of sea stars in the Pacific from a condition known as sea star wasting disease, which is linked to warming waters. With the loss of sea stars, which are one of the main predators of purple sea urchins, the urchin population exploded and devoured the kelp forests. "It's very scary," Rogers-Bennett said. "The Blob shows you how fast a tipping point can happen."

In the past decade, scientists have detected marine heat waves around the world: The Mediterranean was hit in 2012, 2015, 2017, and 2022. One Spanish oceanographer called the heat waves in the Mediterranean, where water temperatures spiked as much as eleven degrees above normal, "the equivalent of underwater wildfires, with fauna and flora dying just as if they had been burned." In 2018, a marine heat wave appeared off the coast of New Zealand and helped spike land temperatures to record highs. Along the coast of Tasmania, giant kelp once stretched over 9 million square meters. Today, thanks to warmer water and a subsequent invasion of urchins, the kelp covers fewer than 500,000 meters. In 2021, a blob of hot water bubbled up in 130,000 square miles of ocean off the Uruguayan coast, an area nearly twice as big as Uruguay itself. It caused a die-off of clams and mussels, an important food source for tens of thousands of people who live on the coast.

These marine heat waves are driving a massive reorganization of underwater life, with many creatures migrating to cooler waters. "Right now, you can go diving off the Monterey pier and see spiny lobsters," said Van Houtan. "They are a subtropical species that are normally found down in Baja. It's absurd to see them up here." (Less absurd, but considerably more dangerous, is the fact that warmer waters are also encouraging juvenile great white sharks to linger in the area.) At the Bodega Marine Lab, scientists documented thirty-seven species that had never been found so far north before. Bull sharks have been hanging out off the coast of North Carolina, five hundred miles north of their habitat in Florida. Lobsters have all but vanished from Long Island Sound. Warm-water species like longfin squid and black sea bass are turning up in the once-chilly waters of the Gulf of Maine. These migrations are radically changing underwater ecosystems, as well as the lives of people who depend on healthy fisheries. Nations in the tropics will likely be hit hardest by fish migration. By 2100, some countries in northwest Africa could lose half their stocks as fish move to colder water. "If you know you are losing a stock, then the short-term incentive is to overfish it," said James Salzman, a professor of environmental law at UC Santa Barbara. "What have you got to lose? The stock's going to move anyway."

Marine heat waves are also inflicting massive damage on coral reefs (where they are often called bleaching events). Reefs are the most biodiverse ecosystems on the planet — they occupy less than 1 percent of the ocean floor, but are home to more than 25 percent of marine life. Reefs are created by millions of coral colonies that build calcium carbonate skeletons. For the past hundred million years or so, corals have thrived in a happy marriage with microscopic plants called zooxanthellae that live embedded in their tissues. Zooxanthellae produce 85 to 95 percent of corals' food through photosynthesis. In return, corals give the plants protection, nutrients, and CO_2, one of the ingredients for photosynthetic food production. This marriage, however, is exquisitely sensitive to changes in ocean temperature. One or two degrees of warming, and the zooxanthellae become toxic to the corals. The corals spit them out and eventually starve to death, leaving only their bleached skeletons behind.

Australia's Great Barrier Reef, a UNESCO

World Heritage site and one of the crown jewels of the natural world, has been hit hard by warming. The reef stretches about fourteen thousand miles along the east coast of Australia — it's the largest structure built by living organisms on the planet, so big it's visible from space. The Great Barrier Reef also supports a vibrant tourism industry, worth $4 billion each year, that employs 65,000 people.

I have visited the Great Barrier Reef twice, once in 2011 and again in 2018, and have seen the decline in real-time. On my second trip, several coral gardens and bommies I had visited on my first trip and remembered as underwater carnivals of color and life were now ghostly, haunted by lonely parrot fish and rays. Since 1998, the reef has suffered six bleaching events, including devastating back-to-back heat waves in 2016 and 2017. Further bleaching in 2020 and 2022 has scientists worried it will be a near-annual event. According to Terry Hughes, a marine scientist at James Cook University in Queensland, Australia, 93 percent of the corals in the Great Barrier Reef have been impacted by some level of bleaching. "We've now added enough greenhouse gases to the atmosphere that mass bleaching of the reef is at risk every summer," Hughes said. "It's like

Russian roulette." In 2015, the Australian government announced an initiative called the Reef 2050 Plan to improve the quality of the reef by reducing pesticide and nitrogen runoff from nearby agricultural regions. However, the plan has been widely criticized for ignoring Australia's contribution to climate change. As Hughes wrote in a 2021 editorial: "Australia has not acknowledged the obvious link between its responsibilities for managing the Great Barrier Reef for future generations, and the damage caused by its ongoing promotion of fossil fuels." Indeed, on my visits to the Great Barrier Reef, I've disembarked from the mainland at a dock only a stone's throw from a massive coal transport terminal, where coal-filled barges carry Australian coal to Indonesia and China. Watching a coal barge float over the Great Barrier Reef is a surreal sight for anyone who cares about the future of the reef — or life on our planet.

Scientists are exploring various ways to make individual corals more heat-tolerant, including creating superstrains that are better adapted to a warmer world. One experiment led by microbiologist Raquel Peixoto from Saudi Arabia's King Abdullah University showed that lathering corals in probiotics could improve coral survival after a heat wave by 40 percent. Peixoto is experimenting with robotic submarines that could drop slow-release probiotic pills onto reefs to release bacteria slowly over weeks.

Some reefs are proving to be more resilient to heat than others. The so-called Coral Triangle of the East Asian Seas is an example. This region holds almost a third of the world's coral reefs — and it is arguably healthier today than in the early 1980s, despite rising water temperatures. That's thought to be thanks to genetic diversity among the region's six hundred species of coral, which is allowing corals to adapt to warm waters. In the Red Sea, corals have evolved in warm, super salty waters, giving researchers hope that at least some varieties of Red Sea coral might thrive in a hotter world.

Still, even the healthiest reefs can only adapt so fast. "By midcentury, pretty much every reef in the world will be eroding away," Ken Caldeira told me. That's astonishing. Coral reefs have been around for about 250 million years, evolving into some of the most complex, diverse, and beautiful living structures on Earth. And yet, if nothing changes, within forty or fifty years they will be crumbling ruins. "I think if we stopped emitting CO_2 tomorrow, some reefs would probably survive," Caldeira said. "But if we go on a few more decades, I think the reefs are gone. Over geological time scales, they will come back, depending how long it takes the ocean chemistry to recover. But it's likely to be at least ten thousand years before anyone sees a reef again."

8. The Sweat Economy

On the first day of the Pacific Northwest heat wave, Sebastian Perez worked alone in a field at Ernst Nursery & Farms in Oregon's Willamette Valley, about thirty-five miles south of Portland. He had just turned thirty-eight years old. Perez has dark eyes and a strong, sturdy build. He wore jeans, work boots, a long-sleeve cotton shirt, and a beige SwissTech safari-style hat that you could pick up at Walmart for $12. He had arrived in Oregon two months earlier, in April 2021, after a treacherous journey from his home in Guatemala, through Mexico, and across the US border. "He wanted to earn money to build us a little house in Guatemala," his wife, Maria, told me. "That was our dream."

On Saturday, June 26, Perez had started work at 6 a.m., with only a brief break at lunch, dragging thirty-pound irrigation pipes between rows of young trees so they would have enough water to ride out the heat wave everyone knew was coming. By noon, the sun was a shimmering ball of fire. Perez knew it was going to be a hot day, but he had no idea what that really meant. This was Oregon, after all, not Death Valley.

The day kept getting hotter: 102 degrees, 103 degrees, 104 degrees. By early afternoon, Perez's heart pounded. Veins swelled in his arms and hands. He felt light-headed, a little woozy. Maybe he took a few swigs from his plastic jug half full of water — after sitting in the sun for hours, the water would have been almost hot enough to scald him. Maybe he looked across the field at the shady stand of Douglas fir in the distance and wondered if he could dare take a break. All around him were rows of young evergreen arborvitaes and boxwoods, two decorative trees grown by the nursery that are popular with Home Depot shoppers. They are plants of suburban sprawl, useful as green barriers around parking lots and for filling in dead zones around the drive-up window at Starbucks, but too small to offer protection from the sun. In the east, Mount Hood loomed on the horizon, an active volcano not unlike the volcanoes where Perez grew up and was eager to return to. He messaged or spoke to Maria, who was still in Guatemala, several times a day and missed her more and more the longer they were separated.

At some point, with his heart pushing hot blood through his body and his hat soaked with sweat, Perez must have known he was in trouble. But he kept working. It was what he had come to America to do, and he never expected it would be easy.

At about 3 p.m., as the temperature hit 106 degrees and kept climbing, the guys working in other parts of the nursery knocked off for the day. They met under a tree and assumed Perez would join them soon. They drank water, they sweated, they waited. They called Perez's cell phone, but no answer, which was strange, because he always kept his phone handy. Finally, they decided to go look for him.

It took them a while to find him. He was lying in the field, among the variegated boxwood, barely breathing. They gave him water, but it didn't help. They dragged him over to the thin shade of a Douglas fir at the edge of the field. He was unconscious by then.

At 3:37 p.m., one of the workers dialed 911. The call came into St. Paul's fire department, roughly five miles away. The worker had a hard time communicating exactly where they were — he didn't speak English well and was unfamiliar with the names of the roads nearby — so it took the ambulance longer than it should have to get to the location. By then, Perez had stopped breathing.

My grandfather and my father were both landscape contractors in the San Francisco Bay Area. They built parks and landscaped highways, schools, and commercial buildings. I worked for them during summers when I was a teenager. All the work was outside. I drove dump trucks, planted trees, poured concrete with other workers, many of whom were immigrants from Mexico or Asia. We sweated a lot. But no one died. In fact, the idea that outdoor temperatures could be deadly never entered my mind (nor, I'm guessing, did it enter the minds of the people I worked with). I loved working outside. Fresh air, sunshine, the dusty shade of California live oaks.

But it's different now. For one thing, it's hotter outside than it used to be. For another, our economy has changed. More and more people work indoors, where the air is filtered and the sunlight is tamed. But not everyone. Somebody still has to build houses, grow food, fix roads, deliver your packages. In the US, fifteen million people have jobs in which they spend at least part of the time outdoors. For them, as well as for warehouse and factory workers who labor in poorly designed buildings, heat is a workplace hazard they confront on a daily basis. And it takes a toll. During the 2021 Pacific Northwest heat wave, Kenton Scott Krupp, fifty-one, was found dead at an Oregon Walmart warehouse. Temperatures had reached 97 degrees the day he died, and his coworkers saw him stumbling and struggling to speak. In Hillsboro, Oregon, a roofer died after collapsing from heat stress on the job.

Postal workers and delivery drivers are particularly at risk, in part because their delivery vehicles often aren't air-conditioned and heat up like convection ovens. In 2021, Jose Cruz Rodriguez Jr., a twenty-three-year-old UPS driver, was found dead in the company's parking lot in Waco, Texas, just days after starting the job. In 2022, twenty-four-year-old UPS driver Esteban David Chavez Jr. died while delivering packages in California. Another UPS driver was captured on video stumbling and collapsing outside a home in Arizona. "People are dropping like flies out here," one UPS driver told the *New York Times*. "It's very brutal." UPS workers shared photographs on social media with thermometer readings of 150 degrees in the backs of their trucks.

If it's brutal in Arizona, imagine what it's like in Qatar, where tens of thousands of migrant workers from Nepal, India, Bangladesh and other countries labored to build new stadiums and hotels ahead of the 2022 FIFA World Cup competition, which was held in the gulf state. With summer temperatures as high as 113 degrees, workers were prohibited from working outdoors in unshaded areas from 11:30 a.m. to 3 p.m. Still, hundreds of workers — or, more likely, thousands — died from heat exposure (officials in Qatar were in no hurry to launch an investigation into migrant worker deaths).

"Rich countries have brought much of their economy indoors where air-conditioning is a possibility, but many developing economies rely on labor-intensive outdoor work," says climate scientist Ken Caldeira. "The combination of poverty and extreme temperatures can be lethal."

Heat is especially dangerous for farmworkers. A 2015 study found that they are thirty-five times more likely to die in a heat-related death than those in other occupations. A United Farm Workers poll of 2,176 farmworkers in Washington, which is not a state known for broiling summers, found that 40 percent had experienced at least one symptom associated with a heat illness while at work. A quarter said they did not have enough cool drinking water, and 97 percent said they thought work protections for heat should be improved in the state.

In addition to the immediate risks of heat exposure such as heatstroke, there can be serious long-term health consequences. In El Salvador and Costa Rica, an epidemic of chronic kidney disease has hit farmworkers who work in hot sugarcane fields — twenty thousand workers have died since 2002 and thousands of others have had to go on kidney dialysis to survive. The disease has been rising among workers in hot climates around the world, including Florida and California. An editorial in the *New England Journal of Medicine* predicts that chronic kidney disease "is likely to be just one of many heat-sensitive illnesses that will be unmasked and accelerated by climate change."

And when it's hot, workers make more mistakes and are injured more often. Researchers at the University of California, Los Angeles, found that even a modest rise in temperature led to twenty thousand additional injuries per year in the state, with a social cost of $1 billion. When it's hot, workers need more breaks, their cognitive skills slow down, equipment fails. Another study found that worker productivity losses from extreme heat in the US totaled $100 billion in 2020 and could grow to $500 billion by 2050. Worldwide, the economic impacts will be felt most powerfully in the Global South. In Dhaka, Bangladesh, for example, which has a labor-intensive economy and where few workers have air-conditioning, heat and humidity already cause productivity losses of about $6 billion each year. The people who take the hit, as always, are those who can least afford it: sidewalk vendors, garment industry workers, and brickmakers who tend the thousands of hot, soot-belching brick kilns in the city.

In the US, there are no federal rules related to heat exposure for workers — indoors or out. Farmworkers, who are excluded from national laws requiring overtime pay, as well as the right to collective bargaining, are particularly vulnerable. For decades, farmworker groups and labor activists have been lobbying the Department of Labor's Occupational Health and Safety Administration (OSHA), which is responsible for workplace safety and worker rights, to develop rules for heat exposure. In 2021, the Asunción Valdivia Heat Illness and Fatality Prevention Act was introduced in the US House of Representatives. The main purpose of the legislation, named after a California farmworker who died of heatstroke in 2004 after picking grapes for ten straight hours on a 105-degree day, is to finally require OSHA to develop heat rules. As I write this in early 2023, the chances that the legislation will make it to the House floor for a vote anytime soon are virtually zero.

State regulations aren't much better. At the time of Sebastian Perez's death in 2021, only California and Washington had rules in place for outdoor workers. The California rule, which passed after several farmworkers died of heat exposure in the Central Valley, requires employers to provide enough fresh, cool water nearby so that workers can drink one quart an hour, to encourage workers to take five-minute breaks every hour, and, most important, to provide shade whenever the temperature rises above 80 degrees.

But in Oregon, farmworker advocates had been fighting for heat rules for nearly a decade, and had succeeded only in getting a few weak directives issued from the Oregon OSHA. "Whenever we tried to get growers to pay attention to the risks of heat in workers, they always talked about how they couldn't afford it," one Oregon organizer told me. "They said prices are set by retail buyers, and changes in labor costs would make them uncompetitive — which is, of course, complete bullshit."

Sebastian Perez grew up in the northern Guatemalan city of Ixcán, on a small farm with his mother and father. In the early 1980s, while a bloody civil war was raging in Guatemala, Perez's parents took refuge in Chiapas, Mexico, where Perez was born. But his family returned to Ixcán when he was young and began farming on a small plot of land. The region was still recovering from the long civil war, and life was tough. By the time Perez was in his early twenties, he was making trips to the US to find decent-paying work to help support his family and pay for medical costs for his father, who was ill. When he was twenty-nine, his cousin introduced him to a gentle woman with a shy smile named Maria. They fell in love and got married soon after.

"I'd go with him into the fields sometimes," Maria told me during a phone call from Guatemala. "We'd hang out together working, so he could finish work faster. Afterwards, when I needed a bath, he would get water for the tub, often making two or three trips to the well to get enough. In the kitchen, when I was cooking, he would help peel the corn. That is how we lived, and we were very happy together."

Still, Perez knew it would be extremely difficult for them to build a future together working on farms in Ixcán. They shared a house with family on a tiny plot of land with no running water. Perez wanted his own house, his own family.

Perez's nephew Pedro Lucas had made the trip to the US a decade earlier. It had been a hard crossing — sixteen days in the Chihuahuan desert, where he fell and broke his knee and was eating plant roots to survive — but he made it. Lucas found a good job at a nursery in Oregon, paying $14 an hour. Lucas's father, brother, and cousin all followed and found work in the Willamette Valley. It was not an easy life, but it was better than the $6 a day you could earn on a farm in Guatemala.

Perez didn't want to move to the US, but he wanted to earn the money to build a house. He also wanted to have children, and Maria needed surgery for a cyst on one of her ovaries if she was going to have any chance of getting pregnant. So in April, Perez decided to go. The journey would be impossible without a guide, known as a coyote. It wasn't cheap: $12,000, cash only. Perez had no money, so Lucas's father agreed to put the title for his own house in escrow to secure a loan. It was a lot of money, but Perez was confident he could work hard in the US and pay it off. Perez crossed the Rio Grande at night just west of Big Bend Ranch State Park in Texas, where the river is narrow enough to easily swim across, then made a brutal trek across the hot desert. He was picked up by a truck outside of Marfa, a little town in south Texas that has been colonized by artists and hipsters, and eventually made his way up to Oregon, where the pay was good and the weather was nice.

When Perez arrived in Oregon, he had few possessions beyond his cell phone, a toothbrush, and the clothes on his back. Lucas set him up in the upstairs bedroom of his house in Gervais, a small community in the heart of the Willamette Valley. The room was bare except for a wood-frame twin bed with a soiled mattress and curtains made of red bedsheets. Within a few days, Perez was working alongside Lucas at Ernst Nursery.

Like every undocumented migrant worker, Perez lived in constant fear of being deported. He had just wagered a $12,000 bet that he could come to the US and make something of himself. It was an enormous gamble. At the rate he was being paid, it would take five months of work in the fields just to pay off the debt. And if he got caught in an immigration raid and deported, that $12,000 loan was still due.

The financial facts of Perez's life were merciless: He worked ten hours a day, at a rate of $14 per hour (low as it is, it's much better than the $8 per hour farmworkers are paid in Texas). No overtime, no paid days off, and for damn sure no health care. After taxes, Perez took home between $2,000 and $2,400 a month (virtually all undocumented workers pay taxes, even though few ever claim Social Security or Medicare benefits later). He paid $500 a month rent for his room at Lucas's house. Whatever was left over he sent back to his mother or to Maria in Guatemala, or used to pay off the coyote loan. According to Lucas, Perez had already paid about $3,000 on his loan, which still left him about $9,000 in debt. "His goal was to pay it all off by December," Lucas says.

Beyond work, Perez's life was exceedingly simple. According to Maria, he was in good health and was taking no medications. He played chess and checkers after work with Lucas in the living room, where a Guatemalan flag was tacked to the wall beside a small flea-market-style painting of Mount Hood. He talked about the house he wanted to build. He talked about how much he missed

Maria. He talked about his mom, whom he adored and called every day (his father had died eight years earlier).

"Sebastian was a serious guy," Lucas told me. "He didn't joke around much. He didn't drink. He didn't party. He went to church. He mostly talked about his future, and the life he and Maria would have together when he got back to Guatemala."

The night before he collapsed in the field, Perez had called both his wife and mother in Guatemala. His mother was concerned. "It's going to be hot," she told him. "You have to be careful." Maria also knew the heat was coming, but told me she wasn't worried. "When it's hot, Sebastian knows to take a break," she said. "When he was working in the field in Guatemala on a hot day, he would always quit work and come home and cool off. Why wouldn't he do that in America too?"

As I write this chapter at home in Austin, a man named Jose is working in the backyard. It's a mild spring morning, and he's digging up a sage plant and moving it to a new spot with better sun. Jose, who grew up in Guerrero, Mexico, is a little older than Perez was when he died, and, like Perez, crossed the border in search of a better future for himself and his family. Jose has a round face, kind, dark eyes, and the weathered body of a man who has worked outdoors all his life. He sometimes brings his twelve-year-old stepson, whom he clearly dotes on, to work with him. My wife, Simone, has known Jose for fifteen years. She met him through a friend when she was looking for someone to help her with her garden. He's been working with her one or two weekends a month ever since.

According to Jose, pruning plants beneath the shade of the oak trees in our yard is a far cry from his real job, which is on a highway crew. Here in Texas that means he is working on black asphalt, which is warm even on cool fall days and a frying pan on hot summer days. When I ask Jose about it, he says that it is hard but that he knows how to take care of himself. He takes water breaks. He eats lunch in the shade in the trucks. The hotter it gets, the slower he works. And when the heat gets too extreme, they stop for the day. This is a man who is skilled in the ways of heat, and that makes all the difference. It's the new guys who get into trouble, Jose says. The guys who ignore it, or who have never experienced it before, or who think they are tougher than the heat.

One of the things you notice living in Texas

is that most of the hard, hot work — in the fields, on the highways, on the building sites — is done by Mexicans. Or Central Americans. Part of this has to do with the proximity to the border, of course. There are a lot of people from Mexico and Central America here looking for work, and some don't have the papers or the qualifications for a job in an air-conditioned office. Another part of it is the fact that they are willing to do work that others aren't, like install a new roof on a sweltering summer day. But it's also hard not to see the legacy of racism in this. It's usually not overt. It's more often expressed in some version of this: Mexicans are from a hot place, so working in the heat doesn't bother them as much. They're used to it.

This is false on several levels. For one thing, Mexico is a big, diverse country, with a vast range of ecosystems, from high mountains to breezy beaches. All Mexicans are no more "used to" hot weather than all Californians.

More important, the idea that people who live in hot places are fundamentally better suited to heat than people who live in cold places is not true — at least, not in a simple and straightforward way. Some people have more-active sweat glands than others, but that is because of complex physiological processes and genetic variation that have nothing to do with where you live or your race. Neither does the color of your skin make much difference. "Infrared radiation is primarily responsible for the heat build-up caused by sunshine, and dark skin and light skin absorb infrared radiation from the sun to nearly equal extents," writes anthropologist Nina Jablonski in *Skin: A Natural History*. "By far the most significant factors in increasing a person's heat load are external temperature, humidity, and the amount of heat the person generates as a result of exercise."

In the eighteenth and nineteenth centuries, heat and racism were deeply entwined. The belief that some races (i.e., nonwhite) are better suited to heat than others (i.e., whites) was widely held, especially in the antebellum South. It helped provide a moral justification for slavery and allowed slaveowners to ignore the horrific working conditions that slaves were subjected to in the cotton fields.

And they were horrific. In 1875, one abolitionist recalled that when the slaves "scraped" the cotton, they were "compelled to go across a thirty, forty, or fifty acre field without straightening themselves one minute, and with the burning sun striking their head and back, and the heat reflected upwards from the soil onto their faces." Making it to the end of the row, where one might briefly stand straight up and perhaps drink some water, took between an hour and an hour and a half. Some slaves, the abolitionist noted, "could not stand straight to save their lives from constant stooping," their bodies bent in forced tribute to the cotton plant.

Samuel Cartwright, a Louisiana slave doctor and racial theorist, argued that African Americans didn't suffer in these conditions because they were different than white people. In the 1850s, Cartwright became the foremost Southern proponent of a brand of scientific racism that made heat tolerance one of the defining differences between the plantation owners and their human chattel. "The practice of negroes in exposing their bare heads and backs, through choice, to the rays of a sun hot enough to blister the skin of a white man proves that they are under different physiological laws from him," Cartwright wrote to US Secretary of State Daniel Webster in 1851. In his view, African Americans toiling in the heat was the natural order of things, and one upon which the entire economy of the South depended: "The labor requiring exposure to a mid-day's summer sun, from the laws of the white man's nature, cannot be performed in the cotton and sugar region without exposing him to disease and death; yet the same kind of labor experience proves to be only a wholesome and beneficial exercise to the negro."

Cartwright was far from alone in this view. In 1826, South Carolina doctor and planter Philip Tidyman argued that the African American was "protected by the very nature of his constitution from the unhealthiness of hot climates, which are so inimical to the whites, especially among those who may be necessitated to labour in low swampy situations." Under such difficult conditions, Tidyman had seen slaves "working with cheerfulness and alacrity, when the white labourer would become languid and sink from the effects of the torrid sun."

A Southern doctor named William Holcombe suggested that the heat tolerance of slaves was due to an anatomical distinction: "The skull of the negro is very thick, dense and strong, resisting injuries and the effect of the heat to a wonderful degree." Holcombe concluded that racial peculiarities "prove the negro to be organically constituted to be an agricultural laborer in tropical climates — a strong animal machine." New York doctor John Van Evrie agreed: "[The negro's] head is protected from the rays of a vertical sun by a dense mat of woolly hair, wholly impervious to its fiercest heats, while his entire surface, studded with innumerable sebaceous glands, forming a complete excretory system, relieves him from all those climatic influences so fatal, under the same circumstances, to the sensitive and highly organized white man. Instead of seeking to shelter himself from the burning sun of the tropics, he courts it, enjoys it, delights in its fiercest heats." To Van Evrie, a divine design explained the slave's constitution: "God has adapted him, both in his physical and mental structure, to the tropics."

Even after slavery was abolished and scientific racism was exposed as a manifestation of white privilege and stupidity, the idea that some races were better suited to heat than others persisted. In 1908, a federal study noted an influx of Mexicans into less-skilled jobs across the Southwest, especially in construction and agriculture. One investigator observed that "they work well and are contented in the desert, where Europeans and Orientals either become dissatisfied or prove unable to withstand the climate."

Labor shortages during World War I drew large numbers of Mexicans and Central Americans into Northern industrial centers.

There, management channeled them into the hot jobs for which they were presumed to be natural candidates. In Detroit, Mexicans worked in the foundry departments of auto plants. In Chicago, they sweated at the mouths of coke ovens and blast furnaces. In 1925, a Pittsburgh pipe manufacturer codified the stereotypes for thirty-six groups on what it termed "Racial Adaptability to Various Types of Work." Mexicans received a positive rating only for jobs with hot-dry conditions and those considered dirty. When economist Paul Taylor visited Pennsylvania in the late 1920s, he was struck by the concentration of immigrants in the hottest operations of the Bethlehem Steel mill. "The Mexicans," Taylor reported, "were said to endure heat well."

In the 1920s, the US Congress tried to grapple with all this during debates over immigration policy. In a hearing of the House Committee on Immigration and Naturalization, Colorado congressman William Vaile summed up the problem from the point of view of rich white farm and factory owners: "The South was not fitted properly for manual labor by white men. It is a hot country and the white man does not thrive under manual labor in the South." Texas congressman Carlos Bee, whose father had been a general in the Confederacy, reassured his colleagues that despite what they might have heard about the difficulty and danger of working in hot fields and factories, the Mexican "is a hot-weather plant . . . He does not want to go into a cold climate; he lives in a tropical climate, and he is willing to live in Texas in the summer time."

Ernst Nursery & Farms, where Perez worked, employs forty to fifty farmworkers, depending on the season, virtually all of them undocumented migrants, according to several workers I spoke with. Ernst was flagged by Oregon OSHA in 2007 and 2010 for repeatedly failing to post information about pesticides used at the worksite, as well as the locations and contact information of local emergency medical care facilities. In 2014, the nursery was cited for failing to provide water to its workers. Ernst would not comment on their labor policies at the nursery, but the workers I talked to said the nursery was no better or worse than others in the area: a ten-minute break every few hours, a half hour for lunch, a Porta Potti out in the field. But no special provisions for water and no tents or other structures for shade. In fact, during a reporting trip to the Willamette Valley, when temperatures were in the midnineties, I didn't see any shade structures for workers in the fields of any of the nurseries or farms I visited.

It's not clear how much attention growers even pay to their workers, other than scrutinizing the numbers on their paychecks. Like many other nurseries and farms, Ernst subbed out the task of finding and hiring workers to an independent labor contractor. In effect, independent contractors allow farm and nursery owners to distance themselves from responsibility for who is working on their farm or how they are treated.

It also makes it harder for workers to speak up against unsafe working conditions, since it is often not clear who is in charge, the nursery owner or the contractor. Sometimes, workers are moved around so much that they become disoriented: "We have seen workers who don't even know where they are, or which farm they are on," says Reyna Lopez, executive director of Pineros Y Campesinos Unidos del Noroeste, a farmworker-rights group in Oregon. When I contacted Ernst Nursery to ask them about working conditions on the farm and Perez's death, the woman I spoke with (she asked that her name not be used) refused to comment on any specifics and would only say, "Our hearts are broken by what happened."

A few weeks after Perez's death, officials in Washington and Oregon finally announced emergency rules for heat for outdoor workers. Oregon's rules now require shade and drinking water whenever the temperature rises above 80 degrees, and, when it jumps above 90 degrees, ten-minute cool-down periods for every two hours of work. This may be an improvement, but it's still not much more than the bare minimum for survival. Had those rules been in place, would they have saved Perez's life? It's impossible to say. It's one thing to have a rule; it's another to enforce it. But the simple truth is that in twenty-first-century America, nobody should be doing physical labor in an open field when it's 107 degrees.

"It's enraging, in a slow and violent way, to think about how heat death is entirely preventable," said Elizabeth Strater, director of strategic campaigns for United Farm Workers. "It doesn't take cutting-edge technologies, or expensive machinery. It takes shade and cool water and rest. That is all. The way that this industry has disrespected and refused to provide this to workers is just criminal."

A week after Perez's death, while his body was still at a funeral home nearby, waiting to be shipped back to Guatemala, I walked up the stairs to the room where he spent his last night. Nothing there but a bed, a lamp, and a vase of flowers wilting in the heat. Lucas had put Perez's few possessions into the casket with him.

"Sometimes I think he is still here," Lucas told me, standing beside the empty bed. "At night, I hear his footsteps." He thought Perez continued to live in the house because he wanted to pay his debt back, and because he wanted to take care of Maria and his mother. Lucas, a strong man with hands like eagle talons, turned away and wept.

The next day, I drove out to Ernst Nursery. I checked the temperature: 97 degrees. I pulled off to the side of the road and opened a gate and walked out onto the field of boxwood and arborvitae where Perez had been working when he died. This was not an accident, or a tragic result of unforeseen circumstances. After all, scientists have known for decades that burning fossil fuels are heating up the atmosphere, and that more intense heat waves are one of the clearest manifestations of life on a superheated planet. And we've known that people like Perez — poor, vulnerable, living outside the air-conditioned bubble of middle-and-upper-class privilege — will be the ones who suffer first and suffer most. But the hard math of the climate crisis is that people like Perez are all too often perceived as expendable. They are one-day media stories and statistics in government reports.

In the field where Perez fell, the young trees looked healthy and green and well tended. I pulled out my phone to take a photo, but I got a warning screen: *iPhone needs to cool down before you can use it.* A few hundred feet away, at the edge of the field beneath a Douglas fir, Perez's wide-brimmed hat and water jug still sat on the ground.

I walked over and picked up his hat — it was stained with sweat.

"He used to call me when he was out in the field sometimes," Maria told me a few days later. "He'd say, 'Mari, I'm working hard here, I'm going to come back soon and build us a house.' That was what he always said. I promised I would wait for him. And now he's coming home in a box."

9. Ice at the End of the World

January 30, 2019

Today, we're supposed to be steaming toward Antarctica. Instead, as I write this morning, the *Nathaniel B. Palmer,* a 308-foot-long icebreaker and research vessel I'm aboard, is still tied up at a dock in Punta Arenas, Chile. On the ship with me are twenty-six scientists and thirty-one crew members and support staff, as well as many millions of dollars' worth of scientific equipment. We departed for Antarctica two nights ago, but we had to return to port because of a problem with the ship's rudder. Divers are in the water now — presumably it will get fixed shortly and we'll be on our way.

The trip I'm about to take is the first expedition in a $50-million, five-year-long joint research project between the National Science Foundation and the British Antarctic Survey designed to explore the risk of sudden collapse of Thwaites Glacier, one of the largest glaciers in West Antarctica. On this trip, scientists will poke and prod Thwaites — aka the Doomsday Glacier* — from every direction, map the ground beneath it, measure changes in ocean currents that are bringing warm water to the base of the glacier, and dig up mud near the front of the glacier to better understand how quickly it has retreated during past warm periods. As chief scientist Rob Larter told us during a science meeting aboard the ship last night, "The question we want to answer is, is the West Antarctic ice sheet on the verge of unstoppable collapse?"

How close West Antarctica is to collapse is one of the most urgent and consequential questions of our time. A stable West Antarctic ice sheet means coastal cities around the world will likely have time to adapt to rising seas. An unstable West Antarctic ice sheet means goodbye Miami — and

* I gave Thwaites this nickname in a 2017 article I wrote about Antarctica, and the phrase has subsequently been widely circulated in the media. It has not, however, been universally embraced by scientists, some of whom find the nickname alarmist. Myself, I find the prospect of the collapse of a glacier that could lead to ten feet of sea-level rise in the not-so-distant future plenty alarming.

virtually every other low-lying coastal city in the world.

But to answer that question, we have to get to Antarctica. Right now, as we wait for the rudder to be fixed, everyone is sorting out their gear, meeting their cabinmates (two people to a cabin, in small bunk beds with railings that can be installed so you aren't thrown out of bed during high seas). The ship has five decks, which are connected through a maze of green steel doors and stairways. Because this is, in part, a US government–sponsored trip, this afternoon we all had to watch videos about sexual harassment and environmental rules in Antarctica (which included tips on how to pick seedpods out of the Velcro on your winter jacket so as to not import any invasive species to the continent). We also practiced getting into the lifeboat and donning our bright-orange immersion suits, which would, in theory, keep us warm for a few hours if we had to abandon ship in the icy waters of the Southern Ocean. We have learned that everything must be strapped down and we have tested seasickness medications and been advised where to puke (toilet or trash can, if possible) when the big seas hit as we cross Drake Passage, the notoriously treacherous open water between South America and Antarctica.

Talk over breakfast in the mess hall this morning was about a storm brewing just west of Drake Passage. But as Larter, a veteran of many Antarctic crossings, put it with a wry smile, "There is always a storm on the way to Antarctica."

Definitions of extreme heat, like definitions of pornography, depend a lot on context. Where I live in Texas, a temperature rise of one or two degrees is imperceptible. In Antarctica, however, a change of one or two degrees can be the difference between ice and water, and between stability and collapse. For millions of people in coastal cities around the world, those one or two degrees can make the difference between a nice view of the beach and three feet of water in the living room. Nowhere on the planet does heat have as much leverage over our future as it does in Antarctica.

Antarctica is the size of the United States and Mexico combined, with a permanent population of zero. It is not the territory of any nation, and it has no government, in the conventional sense. Ever since British explorer Robert Falcon Scott and Norwegian Roald Amundsen captivated the world with their race to the South Pole in 1911, it has been a playground for scientists and adventurers (and, in the popular imagination, penguins). Seventy percent of the Earth's water is frozen here in ice sheets that can be nearly three miles thick. The continent is roughly divided by the Transantarctic Mountains; East Antarctica is bigger and colder than West Antarctica, which is far more vulnerable to melting, in part because large parts of the glaciers in West Antarctica lie below sea level, making them vulnerable to melting from small changes in ocean temperatures.

Until recently, climate scientists didn't worry too much about Antarctica. It is, after all, the coldest place on Earth, and except for a small part of the Antarctic Peninsula that juts north, it hadn't been heating up much. It was also thought to be isolated from the warming ocean by a current that surrounds the continent, essentially walling it off from the Atlantic and the Pacific. The United Nation's Intergovernmental Panel on Climate Change Sixth Assessment Report, which was issued in 2021 and is the gold standard for climate change science, projected from 1.2 to 3.2 feet of global sea-level rise by 2100, with very little of it coming from melting ice sheets in Antarctica (although the IPCC did include a caveat suggesting that could change).

The IPCC's sea-level-rise projections have long been controversial, in part because the melting of the Greenland and Antarctica ice sheets is so difficult to predict. A few years ago, NASA climate scientist James Hansen told me that he believed the IPCC estimates were far too conservative and that the waters could rise by as much as ten feet by 2100. For Hansen, the past is prologue. Three million years ago, during the Pliocene era, when the level of CO_2 in the atmosphere was about the same as it is today and temperatures were only slightly warmer, the seas were at least twenty feet higher. That suggests there is a lot of melting to come before the ice sheets reach a happy equilibrium. Mountain glaciers could contribute a little bit, as would the thermal expansion of the oceans as they warmed, but to get twenty feet of sea-level rise, Greenland and Antarctica likely contributed in a big way (as did other, long-vanished glaciers).

For climate scientists, Greenland was an obvious concern. For one thing, the Arctic has been warming up faster than any other place on the planet. For another, the melt there was visible to anyone who cared to look: every summer, as the surface of the ice sheet heats up, water pours off in great blue rivers, some of them falling through holes in the ice called moulins. For scientists, it was a no-brainer. And compared to Antarctica, Greenland is also easy to get to, just a short flight from Europe to one of the old fishing villages on the coast. You can visit the fastest-moving glacier in the world, the Jakobshavn, and be back at your hotel for a whiskey before dinner.

But then things got weird in Antarctica, causing some scientists to look harder at what was going on down south. The first alarming event was the sudden collapse in 2002 of the Larsen B ice shelf, a vast chunk of ice on the Antarctic Peninsula. An ice shelf is like an enormous fingernail that grows off the end of a glacier where it meets the water. The glaciers behind the Larsen B, like many glaciers in both Antarctica and Greenland, are known as marine-terminating glaciers, because large portions of them lie below sea level. The collapse of ice shelves does not in itself contribute to sea-level rise, since they are already floating (just like ice melting in a glass doesn't raise the level of the liquid). But they perform an important role of buttressing, or restraining, the glaciers. After the Larsen B ice shelf vanished, the glaciers that had been behind it started flowing into the sea eight times faster than they had before. "It was like, 'Oh, what is going on here?' " Ted Scambos, a glaciologist at the National Snow and Ice Data Center in Boulder, Colorado, told me in 2017. "It turns out glaciers are much more responsive to heat than anyone thought."

Luckily, the glaciers behind the Larsen B aren't very big, so sea-level rise wasn't a concern. But the Larsen B prompted scientists to look closer at the ice shelves and movement of glaciers elsewhere in Antarctica. Satellite imagery showed that the ice shelves throughout the continent were thinning, especially in West Antarctica. Some were thinning by a lot. It wasn't clear why, since, unlike Greenland, air temperatures in Antarctica weren't warming much, if at all. The only culprit could be the ocean. Scientists figured out that due to changes in the winds and ocean circulation, more warm deep water was being pushed up under the ice shelves, melting them from below. "Just one degree of change is a big deal to a glacier," Scambos told me.

As it turned out, a lot was going on in Antarctica. The ice shelves were thinning, warmer water was pushing in beneath the glaciers, and the glaciers were flowing faster. The whole place was in dramatic flux. How fast would it go? Nobody knew. Was it possible that the biggest threat to submerging coastal cities wasn't Greenland after all, but Antarctica? If all of Greenland were to melt, that's twenty-two feet of sea-level rise. When Antarctica goes, it's two hundred feet.

"Antarctica used to be the sleeping elephant," Mark Serreze, the head of the National Snow and Ice Data Center, told me. "But now the elephant is stirring."

February 2, 2019

We are about two hundred miles off the coast of Chile right now, about to enter Drake Passage. This stretch of water, named after English explorer Sir Francis Drake, has long terrorized sailors who venture into what is called the inhospitable latitudes. The winds often scream through the Drake from the west, swirling around Antarctica with nothing to stop them, giving the winds plenty of time and space to build monstrous waves. In addition, the Antarctic Circumpolar Current, which is the strongest ocean current in the world, five times more powerful than the Gulf Stream, swirls around the continent, amplifying the waves and making navigation difficult. Thirty-foot waves, combined with seventy-knot winds, are not uncommon. In 2017 a buoy just west of the passage recorded a wave sixty feet high, the largest wave ever recorded in the Southern Hemisphere. The

Drake is widely recognized as the most dangerous passage in the world for ships. It's a seascape haunted by generations of broken masts, abandoned ships, and lost sailors.

Today, I feel the swell growing. The *Palmer* rocks with a predictable rhythm that is, so far, not terribly alarming. At lunch today in the ship's mess hall, I sit with some of the marine technicians. Most of them have been through the Drake a dozen times or more. They all agree that if they had to be crossing the Drake on any ship, it would be this one. The *Palmer* has four big diesel engines, and a deep, strong hull built out of ten-by-forty-foot steel plates. The ship is classified as an ABS-A2 icebreaker, which means it is capable of plowing through three feet of ice at three knots.

After lunch, I climb five flights of metal stairs to the bridge to get a view of the sea. The bridge of the *Palmer* is a remarkably tranquil place, spacious and surrounded by windows on three sides. The rumble of the engines is muted — it feels like you are flying. There are mountains of waves all around us, with a swell coming in from the west. Playing quietly on the bridge's sound system are the Doobie Brothers, the Grateful Dead, Pat Benatar. It's an odd juxtaposition: wild seas and classic rock.

Rick Wiemken, the chief mate, is at the helm. He grew up in a small town near Chicago and fell in love with the sea when he sailed around the world with his brother when he was eighteen years old. Now he has two kids and lives in Honolulu and spends half his life on the sea, piloting ships like the *Palmer.*

I ask him how many times he has been across the Drake Passage.

"Oh, eight or nine times," he tells me.

"Have you had many rough crossings?"

"I think the first one was the worst," he said. "We had seventy-knot-per hour winds. It was dark. I was afraid we were going to tip over."

I ask him how high the swell is now. It's hard to tell, with no perspective out there.

"Maybe ten feet," he says. "It will get worse tomorrow."

"How big?"

"Depends on the storms. We're trying to figure that out now. This is a good ship. It can handle whatever comes our way. The trouble is, when the big waves come, and they hit you on the beam. Then you have the risk of rolling. If they get too big, you have to hove-to — which means turning into the wave and facing it head-on. You ride up and over them. That is much more stable."

"What does tomorrow look like?" I ask him.

He smiles. "It will be interesting."

The first person to understand the risks of a sudden collapse of West Antarctica in a rapidly warming world was the eccentric Ohio State glaciologist John Mercer. Mercer, who grew up in a small town in England and was known for carrying out his scientific fieldwork in the nude, first visited Antarctica in the mid-1960s. At that time, scientists were just beginning to understand the link between CO_2 emissions and a warming climate. They knew that ice sheets had grown and retreated in the past and caused sea levels to rise dramatically, but the discovery that ice ages were triggered by small changes in the Earth's orbit meant that ice sheets were much more sensitive than anyone had thought to small changes in the temperature. Ice cores and improved mapping also helped scientists understand that ice sheets were not monolithic blocks, but in fact made up of rivers of ice, each flowing their own way and at their own rate. In the late 1960s, Mercer may have been the first scientist to ask the question that is still central today: How stable is Antarctica in a climate that is being rapidly warmed by fossil fuel consumption?

Mercer was most interested in West Antarctica. As far as anyone knows, no human had ever set foot on the West Antarctica glaciers until the International Geophysical Year, in 1957, a Cold War collaboration of the US and the Soviet Union and other nations to expand the boundaries of scientific exploration. A team of scientists had trekked across the glaciers of West Antarctica, including Thwaites; by drilling ice cores and taking other measurements, they discovered that the ground beneath the ice was on a reverse slope, having been depressed by the weight of the glaciers over millions of years. "Think of it as a giant soup bowl filled with ice," one polar scientist told me.

In the bowl analogy, the edge of these glaciers — the spot where a glacier leaves the land and begins to float — is perched on the lip of the bowl a thousand feet or more below sea level. Scientists call that lip the grounding line. Below the lip, the terrain falls away on a downward slope for hundreds of miles, all the way to the Transantarctic Mountains, which divide East and West Antarctica. At the deepest part of the basin, the ice is about two miles thick. In the 1950s, before anyone thought about global warming, this was considered an interesting insight into the structure of Antarctica, but hardly a discovery of huge consequences.

Then, in 1974, Hans Weertman, a materials scientist at Northwestern University, figured out that these glaciers in West Antarctica were more vulnerable to rapid melting than anyone had previously understood. He coined a term for it: marine ice-sheet instability. Weertman pointed out that warm ocean water could penetrate the grounding line, melting the ice from below. If the melting continued at a rate that was faster than the glacier grew — which is currently the case — the glacier would slip off the grounding line and begin retreating backward down the slope, "like a ball rolling downhill," explained Ian Howat, a glaciologist at Ohio State University. As the glacier becomes grounded in deeper and deeper water, more of the ice is exposed to warming ocean water, which in turn increases the rate of melt. At the same time, parts of the glacier want to float, which places additional stress on the ice, causing it to fracture. As the face of the glacier collapses, or calves, more and more ice falls into the sea. The farther the glacier retreats down the slope, the faster the collapse unfolds. Without quite meaning to, Weertman had discovered a mechanism for catastrophic sea-level rise.

Mercer understood that Weertman's breakthrough had big implications. In a 1978 paper called "West Antarctic Ice Sheet and the CO_2 Greenhouse Effect: A Threat of Disaster," Mercer focused on the floating ice shelves that buttress the West Antarctic glaciers. Because they are thinner and floating in the ocean, as the water warms, they will be the first to go. And when they do, they will not only reduce friction that slows the glaciers' slide into the sea, they will change the balance of the glaciers, causing them to float off the grounding line. And that, in turn, would accelerate their retreat down the slope. Mercer argued that this whole system was much more unstable than anyone had yet realized. "I contend that a major disaster — a rapid five meter [16 foot] rise in sea level, caused by the deglaciation of West Antarctica — may be imminent," he wrote, predicting it would lead to the "submergence of low-lying areas like Florida and the Netherlands."

Mercer didn't know how soon this might happen, but when he made his calculations in the mid-1970s, he predicted that if fossil fuel consumption continued to accelerate, it could begin in fifty years. That is, right about now.

FEBRUARY 3, 2019

It was a rough night. The swell was fifteen to twenty feet, I later learned. As the waves rolled by, the ship tilted back and forth like a giant pendulum. With one swing, my feet rose, the blood rushed to my head. Then with the next swing, my feet fell, until it almost felt like I was standing up. Everything in my cabin was stowed away. But I heard stuff tumbling down the hallway, and in the distance, the crash of something solid and heavy.

At breakfast in the mess hall today, there are a lot of ashen faces. This is now our third day in Drake Passage. From the looks on my shipmate's faces, nobody slept well. Water constantly washes over the portholes, creating the feeling that we are almost underwater. Lars Boehme, forty-four, an oceanographer and ecologist at the University of St. Andrew's, Scotland, is on the trip with us. He jokes, "When you see penguins flying past the portholes, you know you are in trouble."

After breakfast, I climb up to the bridge, where Captain Brandon Bell and one of the mates are at the helm. I notice that the ocean is in a different mood today than it was yesterday — instead of yesterday's furious white-capped waves, it's a broad horizon of long, peaking swells. Each swell heaves the ship to thirty degrees and floods the main deck. As I watch, "Brown Sugar" is playing on the bridge sound system as the swells roll the ship back and forth like a rubber duckie in a bathtub. I secure myself in an observer seat on the port side of the bridge. As we roll, the horizon tilts radically and the bridge dips down toward the sea. At the end of each roll, it almost feels like I could reach out and touch the water.

As the ship rocks, Captain Bell is cool. Oddly enough, he's a fellow Texan. He grew up in north Texas, on a ranch his family has had since the 1800s. He raises commercial cattle and rodeo bulls. He has been through the Drake many times. "This is our stretch of the highway, we know it pretty well," he says. I ask him about the highest wave he's seen in the Drake — "Sixty feet," he says.

I joke with Bell that piloting the ship through big waves like that must be like riding a rodeo bull, and he smiles. "It's just a different kind of ride," he says, and explains that the waves are more spread out, so you go up one side, and then over the top, and down the other.

Today, despite the course correction, the swell is still hitting the ship on the beam, which is causing a big roll. On the wall in the bridge is an instrument that tells you how far the ship is tilting. I watch it roll past thirty degrees. The captain and I lean like we are standing in a fun house. I ask him where he gets worried, and he says thirty-five or forty degrees. "It's not like the ship is going to roll over," he explains. "But there can be a lot of damage, a lot of computers flying around."

In the early afternoon, I retreat into one of the labs on the main deck of the ship. The rolls continue to build. Then one hits that is bigger than the others, and the ship keeps tilting until it feels like we are almost sideways. I grab the desk — it takes all my strength to hold on. Chairs topple over. A fellow journalist sitting a few seats away from me struggles to hang on but loses her grip. Her chair flips over with her in it. Everything that isn't secured flies across the room. Across the hall, in another lab, a freezer door flies open, contents spill out. The ship rocks back the other way — it is impossible to stand up. For the first time, I see real fear in people's eyes.

When I get up to the bridge around 8 p.m., Wiemken is at the helm again. "Quite a ride, huh?" he says, smiling. "I think the worst is over. We squeezed through the storms. We're back on course now." We are close to 60 degrees latitude south, near the edge of the Antarctic Convergence, where the water changes and the temperature drops and we arrive in a different world.

February 12, 2019

Aboard the *Palmer* today, there is much talk about how the fate of the civilized world may be determined by the movement of warm circumpolar deepwater currents under Thwaites Glacier. How far and how fast that warm current is flowing under Thwaites, melting it from below, will broadly determine how quickly the glacier collapses.

The trouble is, measuring warm circumpolar deep water in Antarctica is extremely difficult — especially under the ice. High-tech underwater moorings can help, but they only cover limited areas and don't measure the top thousand feet of the ocean. Remote submersible devices gather megabytes of precise ocean data, but they are expensive and best suited to targeted research projects.

Mother Nature has better ways. As it turns out, seals are terrific research assistants in Antarctica. Weddell and elephant seals, in particular, regularly swim through the same waters scientists want to explore, and they do it all year long, even under thick ice. Why not give them a digital notebook and let them record what they see? And that's pretty much what a seal tag does. Using what amounts to a temperature probe hooked up to a satellite phone, seal tags record where seals swim, how deep they dive, and what the ocean's temperature and salinity are wherever they travel. When a seal surfaces, the tag beams the results up to a satellite, which is then relayed to a central database, giving scientists real-time data that they use to calculate changes in ocean circulation, as well as to better understand the behavior of the seals themselves.

Here on the *Palmer,* Lars Boehme is in charge of seal tagging. Boehme, who grew up in a coastal town in Germany, is a witty, cheerful guy who looks like he always has something more important to do than comb his hair. He first sailed across the Atlantic alone when he was eighteen (he's done it twice) and considered a career as a sail-maker before giving it up for science. On the *Palmer,* Boehme has earned the nickname "the seal whisperer" for his obvious connection with seals and his respect for their intelligence and remarkable capabilities. At dinner this week, he wowed a group of us by explaining how a seal can track fish by using its whiskers to detect the minute disturbances the fish makes in the water as it swims.

February 14, 2019

We escape the *Palmer* on a Zodiac boat and beach on one of the Schaefer Islands, where a Weddell seal is lying on the rocks a few yards away from our landing spot. The Schaefer Islands are less than a mile from the coast of Antarctica, an archipelago of windblown rock and ice. The island we happen to find ourselves on is a tiny spit of land that is clearly beloved by Weddell seals and Adélie penguins. Weddell seals are smaller than elephant seals, and don't dive as deep, but they are still big, impressive animals. This one is a female. She is more than five feet long and Boehme estimates she weighs about six hundred pounds. Her fur is light brown, her eyes closed. She isn't moving. I wonder if she is dead. "She's just taking a nice long nap in the sunshine," Boehme explains. Meanwhile, we humans all have on at least five layers of high-tech winter gear, insulated boots, and several pairs of gloves. And we are still freezing. "When you come to Antarctica," Boehme said, "you feel like an alien, while all the animals are well adapted and comfortable."

A few yards away, a big male elephant seal eyes us. He is about three times the size of the female Weddell seal — Boehme guesses he weighs nearly a ton. When Boehme walks over to check whether this seal might be a candidate for tagging, the seal lifts his head and opens his pink mouth, revealing long fanglike teeth, and roars like a lion. Boehme seems amused. "He's cranky," Boehme says. He determines that the seal has not molted yet (that is, shed its coat and replaced it with new fur) so is unsuitable for tagging.

With that, he focuses on the Weddell seal sunbathing on the rocks. He has already checked her fur — she has molted. But Boehme is concerned that the seal is too close to the water and could make a run for it under light sedation. So instead of using a blow dart to tranquilize her, which injects the drug into the muscle and is slow to take effect, they decide to use a head bag to quickly gain control of her and then use a syringe to inject the tranquilizer intravenously. It is a little more difficult, but in Boehme's view, it is safer for the seal.

Boehme and Bastien Queste, an oceanographer from the UK's University of East Anglia who is assisting him, approach the seal with a heavy green canvas bag with straps on each side. The seal wakes at their approach, her big doglike eyes looking at them curiously. Boehme and Queste try to slip the bag over her head, but she twists away. *Who are these strange upright creatures and what are they trying to do to me?* Unlike in the Arctic, where seals are preyed upon by polar bears and flee any movement, seals in Antarctica have no natural predators. And so they have no fear. But they are also highly intelligent.

Boehme and Queste move with the seal, but she dances away again, surprisingly graceful and fast. They wrestle and dance with her for a few minutes. Then, in one quick move, it is done. The seal is hooded. At first, it is hard not to make a disturbing visual association with a Klansman or executioner. But it is apparent that the seal was not hurt. In fact, she calms down. Boehme gently lowers the seal's head to the ground, and the rest of the team quickly approaches. Gui Bortolotto, a veterinarian and marine mammal ecologist at the University of St. Andrews who is working with Boehme, pulls out a syringe. Boehme finds a spot in the vein that runs along the seal's back and injects the tranquilizer. The seal relaxes immediately. Boehme and his team withdraw to give the tranquilizer a few moments to take effect.

Working quickly, Boehme checks that the tag is functioning properly. Then he and Bortolotto and Queste kneel down beside the seal, measure her length (eight and a half feet), and roll her on her side to measure her girth (about sixty-three inches). They fold back the hood to reveal the seal's head, while keeping her eyes covered. Boehme spreads epoxy on the base of the tag, then sets it on the seal's head, moving with a gentleness that he might use if he were crowning a queen. He positions the tag exactly where he wants it and neatly spreads the epoxy along the edges of the tag, wiping away any excess. Then, quietly, Boehme and the others stand up, remove the bag from the seal's eyes, and step back.

The seal's breathing stutters a bit, which Boehme assures me is normal. Then she slowly opens her eyes and looks around, as if waking from a dream. She doesn't know it, but she now has a plastic box with an antenna glued to her head. It looks ridiculous, almost comical. And a lot of people will say it's depressing to see a beautiful wild animal tampered with like this, even if she is unaware of it and feels no pain. Some people might even be outraged by it. But to me, this connection between seal and scientist is moving, even heroic. Within the next hour, she will slide into the cold Antarctic waters and be off on her journey, diving into the deep, surfacing, and sending data back to Boehme, doing her part to help scientists learn about the risk of ice-sheet collapse. She has, in effect, become a heat researcher.

MARCH 2, 2019

At 5 a.m., at latitude 74 degrees 57.4 S and longitude 106 degrees 12.8 W on the remote coast of West Antarctica, the Doomsday glacier looms up out of the fog and reveals itself to us.

Peter Sheehan, a witty and hardworking twenty-seven-year-old researcher from the University of East Anglia, is one of the first scientists aboard the ship to lay eyes on the enormous glacier. He had just climbed up five flights of stairs from the lab to the bridge of the icebreaker, where he takes sea ice assessments every hour throughout the night. And there it was: a big wall of ice on the starboard side of the ship, looming in the early-morning light. "It was an eerie sight — the blue water, the blue sky, the blue ice. It was all shades of blue."

Sheehan, who had never been to Antarctica before, ran down to get his camera and then put on his coat and walked out to the bow of the ship, where he stood alone, one of the first human beings ever to confront this enormous glacier, whose fate is inextricably linked with the future of civilization.

"Usually I have a science mind, I am

considering how we're going to collect data, but at that moment, it was all the human reaction," Sheehan says. "I was just overwhelmed by the power and beauty of it."

By 7 a.m., virtually the entire science and support team on the *Palmer* — maybe twenty-five people — crowds up onto the bridge. We all have our cameras and iPhones out, snapping pictures as we cruise along the hundred-foot-tall craggy face of Thwaites's ice shelf. The weather is oddly warm and welcoming — the wind calm, the sea tranquil. The *Palmer* is able to get within a few hundred feet of the calving front — an extremely rare thing to be able to do at any glacier, much less a giant like Thwaites, due to the risk of falling ice. The fissures in the ice glow luminous blue. Emperor penguins dive off a nearby ice floe and swim beside the ship, surprisingly fast and graceful, leaping in and out of the water, as if they are welcoming the big ship as a long-lost friend.

The peacefulness of such glacial awesomeness is sublime. It is also spooky, as if this massive ice wall marks the boundary of another dimension of time and space.

I stand on the bridge with Anna Wåhlin, an oceanographer from Sweden, who has been to Antarctica seven times before and thought she had seen plenty of ice in every shape and form. But she is obviously moved by her encounter with Thwaites. "We are the first people to be here looking at this," she said, her eyes scanning the blue wall of ice. "Ever."

Chief scientist Rob Larter walks over and joins us. "It's more chaotic than I expected," he says. He explains that most ice shelves in Antarctica are flat and clean, like slices of wedding cake. In contrast, many sections of the Thwaites ice shelf are a jumbled mess, with big crevasses and sloping shoulders. To Larter, the sloping shoulders of the ice shelf mean there is a lot of melting going on at the base of the ice sheet, which means there is probably a lot of warm circumpolar deep water flowing in below.

We cruise along the face of the glacier for a few hours, mesmerized by the wall of ice. "It's like staring at a fire," says Lars Boehme. "You can keep staring at ice forever." As we sail by, a sonar system in the ship's hull sends out mile-wide pulses of high-pitched sound waves, which the ship's computers will translate into colorful, real-time maps of the troughs and undulations of the ocean floor three thousand feet below.

By evening, the wind has kicked up, and the ship turns away from the face of the glacier to map more of the seabed in the region.

People crash out in their cabins, or lounge around in the labs, eating ice cream bars.

I find myself thinking about life in Austin. The music and the bars, the highways and the traffic, the new buildings downtown and the boats on the lake — all the buzz of civilization, all the life, all the heat. I imagine the molecules in the city all vibrating faster, and those molecules bumping into other molecules, until finally the dancing molecules vibrate all the way down to Antarctica, a place so remote that I am one of the first humans ever to sail these waters. I know that isn't how it works, but I also know it sorta is. The heat we generate in modern life cannot be contained. It is not localized, like the soot that spews out of the 4x4s blasting through the city. The heat of life in Austin — or Bangkok or Rio de Janeiro or Sydney — is the heat of the world. And it touches everything.

Later, I find Sheehan at his desk in the ship's lab, back at work on ocean chemistry data. I ask him to show me some pictures he took early this morning from the bow of the ship. He pulls them up on his Mac and flips through them. In each one, the wall of ice looms out of the blue, terrifying in its beauty. I ask Sheehan what he is thinking about now, looking at them. "For me, it's hard to envisage something so big, so permanent, so vast, to be as fragile as it is," Sheehan says. "We equate size and grandeur with permanence — like you look at a mountain, and you think, *That will always be there.* But looking at Thwaites forces you to realize that is not always the case. This glacier, huge as it is, doesn't have permanence. If we come back next year, it will look completely different." Sheehan pauses, looking again at the photo of the glacier. Above his head, just outside the porthole, the real thing looms in the twilight. "Seeing this glacier makes you realize that things you think will always be there might not be. That's quite a thing to get your head around."

10. The Mosquito Is My Vector

Jennifer Jones spent most of the summer of 2020 at home, as so many did during the first year of the Covid-19 pandemic. Jones, forty-five, lives in Tavernier, a community in the Florida Keys just south of Key Largo, and passed a lot of time in her yard, puttering around with plants. At some point, a mosquito landed on her. That's not unusual in Florida, and Jones doesn't remember this mosquito bite in particular. But it was not a garden-variety backyard mosquito. It was *Aedes aegypti,* an exquisitely designed killing machine that is one of the deadliest animals in human history. By one count, half the people who have ever lived have been killed by mosquito-borne pathogens. *Aedes aegypti,* which first arrived in North America on slave ships in the seventeenth century, is capable of carrying a whole arsenal of dangerous diseases, from yellow fever to Zika.

The mosquito could sense the heat of

Jones's body and smell CO_2 on her breath from more than thirty feet away. It landed on her exposed flesh, likely her arm or lower leg. The mosquito was a female — only females drink blood, which they need to produce their eggs. It worked quickly, knowing, in the genetic coding of its insect brain, that the longer it lingered the less likely it was to survive. First, it spit on Jones's skin to numb it so she wouldn't be alerted to the bite. Then it plunged its syringelike proboscis, which is actually a sheath containing six needles, into Jones's skin. It probed around until it found an ideal place to tap into a blood vessel. Then it inserted two needles, each one serrated like a carving knife, to saw a hole in Jones's flesh. Two more needles pried the hole open, which allowed it to insert what looks like a tiny hypodermic syringe into Jones's blood vessel. And here is the important part: as it sucked out the blood, the mosquito spit its own saliva into Jones's veins, which contains an anticoagulant that prevents the blood from clotting at the puncture site. In this case, it also contained a virus that causes a tropical disease called dengue fever. When its appetite was sated and its belly full of blood, the mosquito flew off.

The word "dengue" most likely comes from the Swahili phrase *Ka-dinga pepo,* meaning "cramplike seizure caused by an evil spirit." Dengue is also known as breakbone fever because when you have it, it feels like your bones are breaking. It has been around for centuries, and is most common in Asia and the Caribbean. According to the World Health Organization (WHO), before 1970, only nine countries had severe dengue epidemics. Since then, it has increased tenfold, making the disease endemic — that is, permanently embedded in the local mosquito population — in a hundred countries. The WHO estimates that 390 million people are infected with the dengue virus each year. As the world warms, making more of the planet comfortable for heat-loving *Aedes aegypti,* the mosquito's Goldilocks Zone will expand northward, and to higher altitudes. By 2080, five billion people, or 60 percent of the world's population, may be at risk for dengue. "The fact is, climate change is going to sicken and kill a lot of people," said Colin Carlson, a biologist at the Center for Global Health Science and Security at Georgetown University. "Mosquito-borne diseases are going to be a big way that happens."

It took a week or so for the virus to do its work. Once in Jones's bloodstream, it latched onto her white blood cells and began replicating. She was watering plants when she felt light-headed, and then developed a fever. "I knew something weird was going on," she told me. Rashes. Pain behind her eyes. And bone-break ache in her joints. "I felt like I was a ninety-nine-year-old lady who had been hit by a truck," she said. In rare cases, dengue can escalate to brain swelling and bleeding, which can be fatal (about ten thousand people a year die from dengue). But Jones was lucky. The pain and fever faded after four or five days, and she was almost recovered when her son called her to his room to point out the red splotches on his skin. As soon as she saw them, she knew: dengue.

As it turned out, the Florida Keys, already hit hard by the coronavirus, was in the middle of a dengue outbreak too.

Heat rearranges the natural world and rewrites disease algorithms on our planet. It creates new opportunities for microbes, opening up fresh biological landscapes for them to explore, turning pathogens into microscopic versions of Ferdinand Magellan, expanding the boundaries of the known world. Heat waves, as well as heat-driven climate events like flooding and drought, have worsened more than half of the hundreds of known infectious diseases in people, including malaria, hantavirus, cholera, and anthrax.

"We have entered a pandemic era," wrote Dr. Anthony Fauci of the National Institute of Allergy and Infectious Diseases (NIAID) in a paper he coauthored with his NIAID colleague David Morens. The paper cites HIV/AIDS, which has so far killed at least thirty-seven million, as well as "unprecedented pandemic explosions" of the past decade. It's a deadly list, starting with the H1N1 "swine" influenza in 2009, chikungunya in 2014, and Zika in 2015. Ebola fever has burned in large parts of Africa for the past six years. In addition, there are seven different known coronaviruses that can infect humans. SARS-CoV spilled over from an animal host, likely a civet cat, in 2002–03, and caused a near pandemic before disappearing. Middle East respiratory syndrome (MERS) coronavirus jumped from camels to people in 2012, but never found a way to spread efficiently among humans and died out quickly. Now we have SARS-CoV-2, the virus that causes Covid-19.

As I write this, the precise origins of Covid-19 remain unclear. The simplest explanation is that the virus emerged from the wilds near southern China, then found residence in horseshoe bats before making the jump to humans. The virus, as of this writing, has infected more than 680 million people and caused nearly seven million deaths around the world. The amount of human suffering this tiny microbe has caused is incalculable: lost loved ones, vanished jobs, broken families, and lingering sickness from a virus that will eventually retreat but will never disappear.

And yet we got lucky. "It could have been much worse," Scott Weaver, scientific director of the Galveston National Laboratory in Texas, one of the top viral-research centers in the country, told me. Compared with other pathogens out there, Covid-19 is relatively docile. It is an easily transmissible virus that is far more deadly than the flu, and has mysterious long-term effects. But it doesn't kill three out of four people it infects, like the Nipah virus. It doesn't cause people to bleed out of their eyes and rectums like Ebola. "Imagine a disease with seventy-five percent case fatality that is equally transmissible," says Stephen Luby, an epidemiologist at Stanford University. "That would be an existential threat to human civilization."

The Covid-19 pandemic is often compared to the 1918 influenza, which killed at least fifty million people globally. But it is perhaps more accurately seen as a preview of what's to come.

Thawing permafrost in the Arctic is releasing pathogens that haven't seen daylight for tens of thousands of years. The Vibrio bacteria that causes cholera, a diarrheal disease that haunted big cities like London and New York in the nineteenth century and still kills tens of thousands each year, thrives in warmer water. An even more deadly strain of the same bacteria, *Vibrio vulnificus,* while rare, has been detected more and more frequently in bays and estuaries on the East Coast, particularly around Chesapeake Bay, as well as in Florida in the aftermath of Hurricane Ian in 2022. *Vibrio vulnificus,* if you happen to eat it in raw or undercooked shellfish, might give you a bad stomachache (in rare cases, it can be fatal). If the bacteria gets in a cut or wound, however, it becomes a flesh-eating horror and kills one in five people who come in contact with it.

But the biggest impact on human health and well-being may be the emergence of new pathogens from animals. Through intensive agriculture, habitat destruction, and rising temperatures, we are forcing creatures to live by the cardinal rule of the climate crisis: adapt or die. For many animals, that means migrating to more hospitable environments.

In one recent study that tracked the movement of four thousand species over the past few decades, as many as 70 percent had moved, almost all of them seeking cooler lands and waters. In Alaska, hunters are discovering parasites from a thousand miles away in southeast Canada alive under the skin of wild birds. "A wild exodus has begun," writes Sonia Shah in *The Next Great Migration.* "It is happening on every continent and in every ocean."

During this wild exodus, these migrating animals are likely to bump into new animals and humans they have never crossed paths with before. Carlson, the Georgetown biologist, calls these events "meet cutes" — random encounters where viruses jump species and new diseases are often born. The vast majority of the new infectious diseases that have emerged in recent decades have come from these zoonotic pathogens, as they are called, with bats, mosquitoes, and ticks being among the most competent carriers of new viruses. When they jump to humans, we get pandemics like Covid-19. What's next? "It's really a roll of the dice," says Raina Plowright, an epidemiologist at Montana State University who studies the emergence of new diseases. An estimated 40,000 viruses lurk in the bodies of mammals, of which a quarter could infect humans. In 2019, Carlson and a colleague created a massive simulation that maps the past, present, and future ranges of 3,100 mammal species, and predicts the likelihood of viral spillovers if those ranges overlap. Even under the optimistic climate scenarios, Carlson estimates that the coming decades will see about 300,000 first encounters between species that normally don't interact, leading to roughly 15,000 spillovers in which viruses enter new hosts. Vineet Menachery, a virologist at the Galveston lab, called the prospect "harrowing."

In 1994, in the small town of Hendra, in the suburbs of Brisbane, Australia, a number of racehorses at one of the stables in town started to get sick. No one knew why. The horses were disoriented, their faces swelled, a bloody froth poured out of their nostrils. One of them was seen banging its head against a concrete wall. Several horses collapsed and died. At about the same time, a man named Vic Rail, who worked at the stable, came down with what he thought was the flu. He ended up in intensive care, where his lungs filled up with fluid. Shortly afterward, he died. Six hundred miles north of Brisbane, another man who lived and worked on a horse farm got a mysterious illness, with seizures, convulsions, and brain swelling before he died, twenty-five days after he was admitted to the hospital. Before the outbreaks ended, seventy horses were sick, and seven humans died who had been in close contact with dead or ill horses.

It took months of sleuthing before scientists figured out what was happening: Giant fruit bats — the Aussies call them flying foxes — likely congregated in fruit trees in a horse pasture. The big bats have been common in that part of Australia for twenty million years. But as the rain forests that were their natural habitat were fragmented by roads, logging, and farms, and their food sources became more and more difficult to locate due to a changing climate, they moved into civilization. They roosted in the trees in the pasture, contaminating the grass with their urine, which was laced with a virus that nobody had ever seen before — it would become known as Hendra virus. It leapt to the horses, which had grazed on the grass, and then to the humans who cared for them. Luckily, Hendra virus was not highly transmissible and was quickly brought under control.

This story is important for two reasons. First, it's a classic "spillover event," and one that echoes the emergence of Covid-19, which likely originated in a horseshoe bat somewhere in southern China, northern Vietnam, or Laos. No one is sure exactly where the jump from bats to humans happened. The virus was first detected in Wuhan, China, in late 2019, but that doesn't necessarily mean that it first infected humans there. One hypothesis is that the virus made the leap to humans while someone was exploring a cave and came in contact with infected guano. That person, or perhaps someone they transmitted it to, then traveled to Wuhan, where the virus spread widely enough to be noticed. A more likely hypothesis is that the virus first jumped to an intermediate host, such as a red fox or common raccoon dog, which was then sold at a wildlife market in Wuhan, where the virus made the leap to humans. (The theory that the virus may have escaped from a Chinese lab has been ruled out by most scientists.) "We may never know exactly where or how this virus first made the jump from bats to people," said Plowright. It took thirty years of detective work to determine that HIV likely emerged in 1908 in Cameroon, during a bloody interaction between a human and a chimpanzee.

The second reason the Hendra virus is

important is that it alerted scientists to just how good bats are at harboring infectious diseases. The list of viruses that have jumped from bats to humans is long and terrifying: Hendra, Marburg, Ebola, rabies (it can be transmitted by dogs, raccoons, and many other mammals, but in the US bats are the main reservoir). Why are bats so good at harboring deadly viruses? For one thing, they have immune systems tolerant of infection that allow them to host a wide variety of viruses without getting sick. They live long lives (up to forty years), giving them plenty of time to spread disease. They are very mobile — some species range thirty miles or so each night in their hunt for food. And more important, as the climate warms, they can relocate. "Climate change is affecting bats in profound ways," said Plowright. "Many bat species are insectivorous, and so climate change has a big impact on their food sources, as well as on their physiological stress and where they live and how they interact with humans."

If the Hendra virus alerted epidemiologists to the link between fruit bats and viruses, that link got weirder in 1998, when the Nipah virus, a close relative of the Hendra virus, showed up in Malaysia. Around the same time, two other viruses originating from bats were detected in Asia and Australia, a sign of a serious leap. "Four viruses to emerge from one host animal is unprecedented," Plowright told me. The question was, Why?

Nipah virus was particularly scary. Nipah is a horrible pathogen, causing fever, brain swelling, and convulsions. Its fatality rate is as high as 75 percent. Of those who survive, one-third have neurological damage. It was initially isolated and identified in 1999 among pig farmers and people who had close contact with pigs in Malaysia and Singapore. Fruit bats hanging in the trees near a piggery dropped fruit infected with saliva, which the pigs ate. Nipah virus caused a relatively mild disease in pigs, but nearly three hundred human cases with more than a hundred deaths were reported. To stop the outbreak, more than a million pigs were slaughtered. Then in 2001, a second outbreak occurred, in Bangladesh. This time, people contracted the virus by drinking date palm sap that had been infected by the bats. Of 248 Nipah virus cases identified in Bangladesh between 2001 and 2014, 82 were caused by person-to-person transmission, and 193 ended in death — a 78 percent fatality rate. "The only thing that prevented Nipah from being a widespread pandemic was that it was not transmitted asymptomatically," said Plowright. "With Nipah, people are only contagious when they already know they have it, which makes the virus much easier to contain."

But viruses mutate, and new strains can emerge. Nipah virus belongs to a family (paramyxoviruses) that includes measles and mumps, both of which spread really well in human populations. Small changes in Nipah could enhance its ability to spread human to human, creating a pandemic with a high mortality rate. "If Nipah did become more transmissible," said Stanford's Stephen Luby, "that would be a really Black Death plague–level concern."

To Plowright, the link between the climate crisis and disease is evident. "These bats are dependent on food collection that is regulated by climate," she explained. "When does a forest flower, and what triggers it to happen? It's not well understood, but it's a whole bunch of factors that come together, like the temperature, the season, the rainfall. Climate is a key factor. Things are changing really quickly. You can imagine a network of food caches across a landscape — some of the bats are moving from one patch to the next; one has flowers and nectar, then they die off, and the bats go to the next patch.

You start taking away those patches, get to a point where there's no food, so they end up in people's yards, or at horse stables, or anywhere food is plentiful."

The more contact these bats have with other animals, as well as people, the more opportunities the viruses they carry have to spill over. "SARS-COV-2 has been a humanitarian disaster," Plowright said. "But can you imagine if it was killing half the people it infected after some period of asymptomatic transmission? That's the risk we are taking here. And the quicker the climate changes, the bigger the risk grows."

In a small, sparsely equipped lab in a down-and-out neighborhood in Houston, Max Vigilant was sorting through a pile of hundreds of dead mosquitoes, looking for the winged terrorist *Aedes aegypti*. Vigilant, fifty-eight, is the head of operations at the Mosquito & Vector Control Division of the Harris County Department of Public Health — basically, he is the head mosquito hunter in what is widely recognized as one of the top mosquito-control operations in the United States. His expertise is hard-won. On the Caribbean island of Dominica, where he was born, he got dengue fever when he was sixteen, sweating through it with a home remedy of lemon water. The experience changed his life, and ever since, he has been working at the intersection of mosquitoes and human health.

A few hours earlier, this pile of now-dead mosquitoes had been buzzing around a Houston neighborhood. Vigilant retrieved them from a trap, tossed them in a freezer at the lab for three minutes ("Doesn't take long!" he joked), and now he was sorting out his catch. Soon these mosquitoes would be ground up and run through a series of tests to determine what, if any, pathogens they contained. There are millions of mosquitoes in Harris County. Every week, a few thousand are ground up to see if anything scary pops out. It's not exactly sophisticated screening, but it's more than most cities do.

Most of the mosquitoes in Vigilant's pile belonged to the *Culex* genus, ordinary backyard mosquitoes that are pretty much everywhere in the South. But Vigilant was looking for something else. He poked through the pile, then plucked one out and held it under a magnifying glass. At first glance, it looked the same as the others. He pointed out the bushy eyebrows, which is one way you distinguish a male from a female (this is a female). "See the white stripes on her abdomen?" he said to me, holding it under a big magnifying glass mounted on the desk. "She looks like she is wearing a white tuxedo."

He held her up like a prize, twisting her around so I could see her from every angle. "That's *Aedes aegypti,*" he said. "She's kind of beautiful, isn't she?"

There are roughly three thousand species of mosquitoes in the world. Of those, only a small percentage are of concern from a public health perspective: *Culex pipiens,* which carries West Nile virus, and *Aedes albopictus,* also known as the Asian Tiger mosquito, which has recently arrived in the US from Asia and can carry dengue and Zika, but does not lust after human blood like *Aedes aegypti.*

Aedes aegypti is an extremely competent vector for dengue and Zika, as well as yellow fever and chikungunya, making it one of the most dangerous animals on Earth. And it is also one of the most companionable (or, as Fauci puts it, *Aedes aegypti* is "uniquely anthropophilic"). It's the Labrador retriever of mosquitoes, happiest when it is living in or near our homes, laying eggs in little puddles of clean, fresh water in a bottle cap or the rim of a planter. And because it thrives in higher temperatures than other mosquitoes, it is well adapted to life on a warming planet.

The impact of rising heat on mosquitoes

is fairly easy to model, in part because mosquitoes are very sensitive to temperature changes and will basically move to stay in their happy zone. And that happy zone is expanding. *Aedes aegypti*–transmitted diseases have increased thirtyfold in the past fifty years because of changes in climate, land use, and population. Mexico City, for example, has always been a few degrees too cold for *Aedes aegypti* to get established. Because of that, the city has always been blissfully free of yellow fever, dengue, and Zika, which have haunted the lowlands of Mexico. But now, as temperatures rise, *Aedes aegypti* is moving in. For the twenty-one million people who live in the city, it's an alarming development. Wherever *Aedes aegypti* turns up, dengue, Zika, and other diseases are sure to follow. You can already see this happening in places like Nepal, which, until recently, was nearly free of mosquito-borne diseases. In 2015, Nepal had 135 cases of dengue. In the first nine months of 2022, there were 28,109 cases.

In other places, the changes in mosquito-borne diseases will be more complex. Malaria killed more than six hundred thousand people in 2020, mostly children in sub-Saharan Africa. The most deadly form of the disease is caused by the parasite

Plasmodium falciparum, which is carried by the *Anopheles gambiae* mosquito, a smaller, less elegant creature than *Aedes aegypti,* and more sensitive to high temperatures. As the planet warms, West Africa is likely to grow too hot for *Anopheles gambiae,* which will shift to higher and cooler regions in eastern and southern Africa. A recent study by Sadie Ryan, a medical geographer at the University of Florida, found that under a high-emissions scenario (which would cause more severe global warming), an additional seventy-six million people could be at risk from exposure to malaria transmission in eastern and southern Africa by the year 2080. At the same time, heat-loving *Aedes aegypti* will move into West Africa, vacated by *Anopheles gambiae,* putting millions of Africans at risk for dengue, Zika, and other diseases.

In Houston, as in most of the South, *Aedes aegypti* is established but less common. The city had its first outbreak of dengue in 2003, and a flare-up of Zika in 2016. Vigilant and other members of Harris County Mosquito Control are constantly on the lookout for *Aedes aegypti,* knowing they are harbingers of doom. Their only real tool for fighting them is to spray insecticides, which they do from the back of pickup trucks whenever there is evidence of a flare-up. But *Aedes aegypti,* as well as other mosquitoes, are developing immunity to many commercial insecticides. "We are losing the war," says Galveston National Laboratory scientific director Scott Weaver. Technological advances, such as genetically engineering mosquitoes to produce female offspring that are infertile, hold some promise in the future. In the first field experiment of its kind, a biotech firm called Oxitec released five million genetically modified *Aedes aegypti* in the Florida Keys in 2021. How effective this strategy will be at reducing the population of wild disease-carrying mosquitoes, however, remains highly speculative. Right now *Aedes aegypti* reigns supreme as the most formidable and insidious vector of future diseases. As Anthony Fauci wrote, "Any virus that can efficiently infect *Aedes aegypti* also has potential access to billions of humans."

The Galveston National Laboratory is a fortress of pathogens, although you would never know it from the outside. It sits on the campus of the University of Texas Medical Branch like any other building. There are some concrete barriers on the outside, and a bunch of weird-looking exhaust systems on the roof, but otherwise, it could easily be the building where you took Chemistry 101 in college. Inside, in one of about a dozen Biosafety Level 4 labs in the United States, scientists work on some of the most lethal viruses in the world: Ebola, Nipah, Marburg, and others.

The BSL-4 lab is Dennis Bente's workroom. A broad-shouldered guy with a full dark beard and a slight German accent, Bente grew up in a small town in northwest Germany and studied veterinary medicine in Hannover before developing an interest in vector-borne diseases. He worked with mosquitoes for a while, then decided ticks were more compelling.

The BSL-4 lab is basically a big concrete box within the larger lab. Entering it is like a journey into deep space. Bente first passes through a buffer corridor, where he grabs a clean pair of scrubs. Then he enters a changing room, where he strips off his street clothes and pulls on the scrubs. Next is the suit room, where he steps into what he calls his space suit, including built-in gloves and a clear plastic helmet. To pressurize the suit, and give himself air to breathe, Bente hooks up to an air hose and inflates like the Michelin Man. If all is well, he steps into the air lock, which is the most important barrier between the deadly pathogens and the outside world. He opens a heavy, airtight submarine door, closes it, walks a few feet, then opens another heavy, airtight submarine door. Finally, he steps into the hot zone.

Inside, he works with a group of ornate-looking ticks that are native to the Mediterranean basin, known as Hyalomma ticks. They are brown, with yellow stripes on their legs, which are much longer than the stubby legs on deer ticks you see in upstate New York. They look almost spidery, which is not surprising — ticks are arachnids, not insects, in the same family as spiders and scorpions. With their long legs, Hyalomma ticks are the speed demons of the tick world. (On YouTube, you can find videos of Hyalommas running after people like tiny lions in pursuit of an antelope.) Unlike many other ticks, Hyalommas are predators. They are one of the few species of ticks that have eyes (the word "Hyalomma" is derived from the Greek words for "glass" and "eye"). Instead of using CO_2 sensors like other ticks to locate a blood meal, Hyalommas sense vibrations in the ground, and watch for shadows, to chase down a nearby human (or livestock, one of their favorite foods).

But Bente is not studying Hyalomma ticks because of their athletic ability or visual acuity. He is studying them because they are the most competent carriers and transmitters of Crimean Congo Hemorrhagic Fever (CCHF) to humans. One way to think about CCHF is that it's basically a slightly less awful version of Ebola. CCHF often starts with high fever, joint pain, and vomiting. Red spots appear on your face and throat. Then by the fourth day, you get severe bruising and nosebleeds, and in many cases, uncontrolled bleeding from other orifices. It lasts for two weeks or so. There is no treatment, no vaccine, no cure. The fatality rate for people with CCHF ranges from about 10 percent to 40 percent.

As far as Bente knows, the only Hyalomma ticks in America are in the Galveston lab. In the wild, they are found in North Africa, Asia, and parts of Europe (in Turkey, there are about seven hundred CCHF cases a year). The ticks, which thrive in warm, dry climates, are expanding their range. In recent years, CCHF has killed people in Spain and northern India.

Bente keeps a colony of Hyalomma ticks in his lab and feeds them on mice and rabbits that he deliberately infects with the CCHF virus. ("The virus has no impact on these animals," Bente pointed out. "It's only dangerous to humans.") He is studying fundamental questions about Hyalomma ticks and CCHF that should freak out anyone who'd like to walk through nature without worrying whether they'll contract a virus that will make their eyeballs bleed: Can Hyalomma ticks be established in the US? (It's extremely unlikely.) Might other types of ticks be carrying CCHF in Africa? (Yes, but so far, they are only "a sideshow," Bente said.) Is airborne transmission of CCHF possible? ("CCHF is a very old virus," Bente said. "Why mutate now?") But Bente still has concerns.

As disease vectors, ticks are very different from mosquitoes. They live up to two years instead of a few weeks. But like mosquitoes, they are sensitive to changes in temperature and can't survive long in cold or dry climates. As the world warms, they are following the heat. Some tick species are moving as much as thirty miles north each year — an unseen parade of bloodsuckers conquering new terrain. They are difficult to target with insecticides, and have many remarkable survival tricks, such as the ability to go long periods without water by basically spitting into a pile of leaves and then drinking it later when they are thirsty. Heat is also changing ticks' appetites. As temperatures rise, brown dog ticks that transmit Rocky Mountain spotted fever — a disease with a 4 percent fatality rate — are twice as likely to choose to bite people over dogs. In the US, ticks can carry more than twenty different pathogens — and more are being discovered all the time. "The more we look at ticks, the more viruses we continue to find," Bobbi Pritt, a microbiologist at the Mayo Clinic in Rochester, Minnesota, told me.

Lyme disease is emblematic of the threat ticks pose in a warming world. It is caused by deer ticks carrying the bacteria *Borrelia burgdorferi.* Lyme was discovered in Connecticut in the mid-1970s. Today it is a major, and growing, health threat. According to the CDC, reported cases in the US have tripled since the late nineties. Lyme disease has become an almost "unparalleled threat to regular American life," as Bennett Nemser, an epidemiologist who manages the Cohen Lyme & Tickborne Disease Initiative at the Steven & Alexandra Cohen Foundation, has said. "Really anyone — regardless of age, gender, political interest, affluence — can touch a piece of grass and get a tick on them."

It's not just the heat that has expanded the range of Lyme-carrying ticks. It's also the increasingly fragmented landscapes in the Northeast. As forests are cut up into suburban developments, the populations of foxes and owls decline, which leads to an explosion in the population of white-footed mice, which are the main reservoir for *Borrelia burgdorferi.* Young larval ticks feed on the infected mice, and then pick up Lyme and later spread it to anyone passing by.

But in Bente's view, the most worrisome development in TickWorld is the invasion of Asian longhorned ticks in the US, which he calls "a cautionary tale." Nobody is quite sure how or when the first Asian longhorned tick (*Haemaphysalis longicornis*) arrived in the continental US. They are native to East Asia, including Australia and New Zealand. They were first reported in 2017, in New Jersey. Within a year, researchers had found the tick in eight other states, and its territory continues to expand as the climate warms and winters grow milder. One key contributor to its rapid spread is the fact that females can reproduce through cloning themselves, without the need for mating, a process called parthenogenesis. This makes it extremely hard to control. "In practice, it's impossible to eradicate this species," said Ilia Rochlin, an entomologist at Rutgers University.

Asian longhorned ticks are aggressive biters, and can gang up on prey to drink large quantities of blood. Their preferred meal is cattle. In parts of New Zealand and Australia, the ticks have reduced production in dairy cattle by 25 percent. So far, there is no evidence that Asian longhorns in North America have transmitted diseases to humans. But that could change. Pritt called the longhorned invasion "extremely worrisome." They can carry several deadly human pathogens, including potentially fatal severe fever with thrombocytopenia syndrome (SFTS) virus and *Rickettsia japonica,* which causes Japanese spotted fever. "While these pathogens have yet to be found in the United States, there is a risk of their future introduction," Pritt told me.

A close cousin of SFTS, as it turns out, is CCHF. What worries Bente is the possibility of what scientists call vector switching. That is, that somehow the CCHF virus jumps from Hyalomma ticks, which are not yet in the US outside of Bente's lab, to Asian longhorned ticks, an aggressive biter that is becoming widespread.

Could CCHF make the leap to Asian longhorned ticks? "Nature is complex," Bente told me. "I don't like the narrative that says we are one tick bite away from catastrophe. But at the same time, I can't say it won't happen."

11. Cheap Cold Air

There have always been a lot of flashy people in Houston. Oil money, Wall Street money, art money, cancer research money — the city swings between boom and bust, attracting outsize characters who thrive on the flash and cash. But Harold Goodman was never part of that Houston. He was of medium build with medium dark eyes and a taste for white shirts and suits that were, at best, medium stylish. His politics were moderately conservative ("He was not a man who was up for a lot of social change," his daughter Betsy Abell told me). He lived in Tanglewood, a quiet but nice Houston neighborhood, with his wife, Harriet, and four kids. His business was air-conditioning, which was definitely not flashy. He was a man who knew how to talk about ducts and cubic feet of air and relative humidity. He played poker once a month. On weekends, he played tennis. He never wore cowboy boots.

There were two glittering exceptions to Goodman's modesty. The first was cars. He loved them. The fact that he was a terrible driver, often hitting the gas and the brakes at the same time, didn't matter. Every year or so, he traded his old one in for something new. His favorite was Lincoln Continentals — over the years, he owned a slew of them. He also owned Porsches, a gold Corvette, a green Oldsmobile convertible, and a particularly sharp coral-red Thunderbird with a white top and white leather seats.

The other exception was horses. He loved to bet on races. It was something that he had done with his father when he was a kid and that continued to be a passion all his life. He read the *Daily Racing Form* and *BloodHorse* and he always knew which horse he liked in the second race at Churchill Downs or the fifth race at Santa Anita. He won some, he lost some, but he never lost the thrill. When he got rich, one of his big indulgences was to buy a seven-hundred-acre horse farm outside of Houston.

Despite Goodman's basic modesty, there must have been a moment when he recognized what he had achieved. Maybe it was when he walked into a Lincoln dealership and felt the whoosh of cold air. Maybe it was when he saw his name on an air-conditioning unit hanging outside someone's window.

At that moment, he must have understood: He was the King of Cool.

I didn't know much about air-conditioning until I moved to Texas. I grew up in Silicon Valley, which was blessed — at least when I was a kid in the 1970s — with a perfect Mediterranean climate. We had no air-conditioning and I didn't know anyone who did. In fact, I never saw or heard of air-conditioning until it turned up in my father's Chevy El Camino sometime in the seventies. I fiddled with it sometimes while my dad drove, but mostly I didn't care. Who needs air-conditioning?

After college, I moved to New York City. I worked as a reporter for a Manhattan weekly in the early 1990s, hanging out with cops and taxi drivers and AIDS activists. During the hot asphalt summers, I discovered why people love air-conditioning. Still, I never had it in any apartment I lived in. When it was hot, I turned on a rotating electric fan and opened the windows and sweated through it. Later I moved upstate to Saratoga Springs and lived in a Victorian house with several fireplaces but no air-conditioning. In Saratoga, I learned that the whole city arose in the nineteenth century largely as a retreat from the heat for Southerners and wealthy city folks, with big trees and big porches that gave people a place to socialize during the hot summer evenings.

When I traveled, I sometimes put up with air-conditioning for a few days. But I hated the way the units rattled and clattered all night long, and the soggy, fetid air in hotels where the windows were sealed shut.

When I moved to Texas, I came to understand heat in a different way. I'd walk out to check the mail and slam into a wall of hot, humid air. Going for a bike ride in the middle of the day felt like a life-threatening adventure. We live in a small house, one that was built before air-conditioning, but which has been retrofitted with a centralized system. In the summer, I sometimes work on the screen porch, beyond the reach of air-conditioning, but the heat is often overwhelming and I retreat inside.

I wonder sometimes how people lived in Texas without air-conditioning. But then I look around and I see the old dogtrot houses with wide breezeways that allowed cooling winds to flow through, the wide porches that gave people a place to sleep on summer nights, and the way houses are built under the shading limbs of big oak trees. I visit public parks like Barton Springs, where I jump into the cold clear water that flows out of the limestone and feel sweet relief. In the world before air-conditioning, there was less CO_2 in the atmosphere to trap the heat, and less asphalt and concrete on the ground to radiate it back at you. It wasn't a better world, necessarily, but it was a different world, and people managed. Playwright Arthur Miller recalled growing up in New York City in the era before air-conditioning: "Broadway had open trolleys with no side walls, in which you at least caught the breeze, hot though it was, so that desperate people, unable to endure their apartments, would simply pay a nickel and ride around aimlessly for a couple of hours to cool off." On Coney Island, Miller wrote, "block after block of beach was so jammed with people that it was barely possible to find a space to sit or to put down your book or your hot dog."

Harold Goodman was born in Beaumont, Texas, in 1926. Oil had been discovered outside the city a few decades earlier — it was the center of the booming Texas oil patch. Goodman's family wasn't part of it — his uncle was a farmer, his father was in the insurance business. The family moved to

Houston not long after Goodman was born, where his father's insurance business prospered. Goodman's sister, Betsy Bramson, who was five years older, told me Harold was "a sharp little boy. But we never thought of him as being really *smart,* if you know what I mean. He loved to play cards and go to the horse races with our father."

In those days, the best cooling device you could find on a hot summer day in Texas came from the Blue Bell creamery in Brenham, a small town about seventy-five miles northwest of Houston. It was a scoop of vanilla ice cream, which old-timers remembered had once been served in a cup out of the back of a horse-drawn wagon. Blue Bell started making ice cream in 1911, using a mix of salt and ice to freeze cream into ice. This was not a new discovery — George Washington had been a big fan of homemade ice cream. But in the early twentieth century, ice cream went from an elitist treat to a democratic pleasure. A scoop of Blue Bell might not stave off heatstroke, but it would make you feel better while you sweated.

For the Goodmans, Houston heat was a fact of life. In the summer, they stayed out of the sun and slept with the windows open or outside on a sleeping porch. "There was no such thing as air-conditioning when we were growing up in Houston," Bramson recalled. "But the funny thing is, I don't remember it being hot. When you're young, you don't pay any attention to it."

Goodman went to the University of Houston for two years, and then transferred to the University of Texas in Austin and graduated with a degree in business. He spent a couple of unremarkable years in the navy. When he got out, he had no idea what he wanted to do with his life. So he went to work for his father in the insurance business in Houston. "He absolutely hated it," his daughter Betsy told me. "He wouldn't show up, or he'd run off and play poker in some basement downtown with his friends at night and get drunk. Everybody in the family was worried about him."

As it happened, his sister's husband had gone into the air-conditioning business, which in the early 1950s was a new but promising venture. "His mother thought that would be good for Harold to do too, but everybody worried because he was not electrically or mechanically inclined." Goodman declined to go into business with his brother-in-law. But another friend was in the business too and finally Goodman decided to give it a try. They started selling window air-conditioning units, which were a new thing then. If Harold Goodman had one talent, it was great timing.

John Gorrie, in contrast, had no such talent. In 1833, Gorrie was a thirty-one-year-old physician living in the frontier swamp of Apalachicola, Florida. Gorrie was a Victorian dandy from Charleston, educated in New York, who found himself trying to cure the scores of people who died of fever in the sweltering wilderness of the Gulf Coast. Like most of his peers in the medical world, Gorrie incorrectly believed that "miasmas," or foul air, generated by decaying vegetation in hot, damp places, were the cause of many diseases, including malaria. If he could cool the air, he believed he could cure malaria.

Gorrie experimented in the fever ward that he maintained for patients in his home. In one test, he suspended a block of ice from the ceiling above a patient. As the ice melted into its container, air was blown onto the block. The cooled air flowed down over the patient and into an opening in the floor. It didn't cure the disease, but it did comfort the patient. The problem was, this method was awkward and dependent on ice shipped down to Florida from New England, which was expensive.

Students of natural science had known for

decades that a gas, compressed, will become hot — and if that gas expands, its temperature will drop sharply. But they had no idea how to put this knowledge to use. Gorrie had an idea. Using a small steam engine to power a mechanism that drew in air, Gorrie built a device that compressed air into a chamber with a piston (the air becoming hot), then forced it into a labyrinth of pipe, where it expanded (becoming cool). The air was then routed through a tank of brine, which itself became chilled below freezing and lowered the temperature of the air even more. "John Gorrie had actually produced a machine that was complete," one historian wrote. "More important, it worked. For the first time, a machine was manufacturing cold air."

This method is called vapor compression, and it's more or less how air conditioners still work today. It is a technology that shaped the twentieth century just as surely as the internal combustion engine did, and has turned out to be just as durable, useful, and troubling for our future.

Gorrie's vapor compression machine worked well enough to comfort the people in his ward, even if it didn't cure them of their fevers. But Gorrie had grander ambitions: he wanted to usher in a new age of cooling. He noted how in northern countries it was customary to design insulated buildings in order to heat them more easily. Now, Gorrie called for the opposite: "Let the houses of warm countries be built with an equal regard to insulation, and a like labor and expence be incurred in moderating the temperature, and lessening the moisture of the internal atmosphere, and the occupants would incur little or no risk from malaria."

But his big ideas went nowhere. In part, it was because his vapor compression machine was big, noisy, expensive, and complicated. But it was also because most people — even friends and neighbors in sweltering Apalachicola — couldn't see the use of it. Sweating was just something people did. The idea of "comfort cooling" didn't exist.

Gorrie later discovered that if he left his machine running long enough, it could manufacture ice from water. So for a few years, he tried to sell it as the world's first ice-making machine, which would save people from the expense and trouble of importing ice from New England. But Frederic Tudor, a New England businessman who had single-handedly built up the ice trade from one tiny ship into a worldwide empire, was not going to let a crazy inventor from Apalachicola kill his business. Tudor made sure Gorrie got zero funding for his ice-making machine. In 1855, Gorrie died broke and broken. His invention was forgotten for the next fifty years.

In 1902, the idea of mechanically cooling the air was resuscitated by a young engineer named Willis Carrier. Carrier arrived at the Sackett-Wilhelms Lithographing and Publishing Company in New York City to help them solve the problem of warping paper in their printing press. As it turned out, high humidity was causing the paper to swell, blurring the printed image as it ran through the press. What could be done about it? Carrier had never thought about the problem of humidity in the air, but, after several tries, came up with a way to "condition" the air by using a fan to blow it over pipes filled with cold water, which caused the moisture in the air to condense on pipes and emerge not just cooler, but drier too. It was an entirely different approach than Gorrie had taken, but the science was the same. Carrier's invention had the added benefit of having a practical business application. The problem with the swollen paper was solved, and the age of air-conditioning was born.

When Goodman started in the air-conditioning business in Houston in the late 1950s, the city was an oil-soaked boomtown.

As one witty character put it, “A fellow can make ten million dollars here if he wants to run after it. A fellow can make one million dollars just by standing still.” In 1940, the city’s population was 384,000. Twenty years later, it was nearly one million — the fastest-growing city in the US.

If oil money was the engine of the boom, air-conditioning was the vehicle. As Fred Hofheinz, the mayor of Houston in the 1970s, put it: “Without air-conditioning, Houston would not have been built at all. It just wouldn’t exist, that’s all.”

The AC invasion began in the cafeteria at the Rice Hotel in 1922. Movie theaters, such as the Texan and Majestic, got chilled in 1926. By 1949, all eleven hundred rooms at the city’s glamorous Shamrock Hotel had air-conditioning. In 1961, the Sharpstown Center became the first air-conditioned, enclosed shopping mall in the US. In 1965, the Astrodome became the first air-conditioned stadium in the world. By 1980, even the most sacred site in Texas — the Alamo in San Antonio — was cooled.

Air-conditioning enabled the building boom not only in Texas, but throughout the South. Goodbye big porches and flow-through ventilation. Hello mass-produced suburban development with cheap construction, low ceilings, and zero airflow. In 1957, the FHA began including the cost of installing central air-conditioning in mortgage loans.

In Texas, it was mostly middle-class white folks who benefited from air-conditioning. By 1960, 30 percent of homes in Texas were air-conditioned, the highest in the country. But only 10 percent of the state's Black population had it.

Nothing captures the rising cultural cachet of air-conditioning like *The Seven Year Itch,* a 1955 comedy directed by Billy Wilder. The film includes one of the most iconic Hollywood images of the twentieth century: Marilyn Monroe standing over a subway grating in Manhattan, her dress blown up around her waist by the rush of air from a passing train.

In the opening scenes of the movie, New York City is in the middle of a scorching heat wave. Women and children are fleeing the city for refuge upstate and at the beach, leaving their hardworking husbands behind to toil in the city. Monroe plays a single woman who is happy to stay and battle the heat. The first time she appears on-screen, she is carrying a big fan up the stairs into her apartment. When her neighbor, a repressed, sex-obsessed family man played by

Tom Ewell, jokes about the heat, she quips, "I keep my underwear in the ice box." The only attractive thing about Ewell is his air-conditioned apartment. He uses it to entice Monroe to visit, boasting that he has air-conditioning in every room. In one scene, Wilder shows Monroe sitting with her long bare legs stretched out in front of an air-conditioning unit. At that moment, AC became sexy.

In the mid-1950s, Goodman started Goodman Manufacturing, making flexible air ducts used in residential air conditioners. The new air ducts, which replaced stiff metal ducts, made it easier and quicker to install air-conditioning in new houses. Business boomed.

A decade or so later, Goodman got into the business of making air-conditioning units themselves. He did it by buying a bankrupt company in Ohio called Janitrol. As Peter Alexander, who would go on to work side by side with Goodman for more than twenty years, recalled: "Harold called me and he said, 'Listen, I'm thinking of buying this company, getting into the manufacturing business. And I'd love to have you join.' I said to him, 'You've lost your mind.' " The air-conditioning business was already dominated by big, established manufacturers like Carrier and Fedders. But Alexander flew down to Houston to talk to Goodman about it anyway. "By the time I was with him for two hours," Alexander recalled, "I said to myself, 'I want to do this.' "

Goodman's first move was to relocate the company to Houston. Why build an air conditioner in the North and then transport it to the South where air-conditioning was king? "When we started the company," Alexander told me, "we said, 'How are we going to differentiate ourselves?' And Harold said, 'We are going to differentiate ourselves under the theory that all the consumer wants is cheap cold air.'

"I know that sounds kind of crude," Alexander continued. "But if you ask ten consumers what brand of residential air conditioners they have in the house, maybe two can tell you. All of these companies that were doing all this advertising — well, what's the benefit of that? So we decided we're not gonna do any advertising. We're just gonna focus on the fact that people don't know what brand is in their house. And that the most important driving factor in their choice of an air conditioner is price. So, we said, 'We're going to be a manufacturer of cheap cold air.' We used to actually say that flat out. *Cheap cold air.* That was our product."

Goodman saved costs by focusing on manufacturing quality to reduce warranty returns and service calls. He kept overhead low. He offshored manufacturing (mostly to Korea). He paid employees by piece rate, a controversial but not illegal labor practice that allowed them to earn more if they worked harder. And most important of all, he pushed the dogma of low price/high-volume sales. "If you sell a hundred units and make seventeen bucks a piece, instead of selling twenty units and make fifty bucks a piece, you're going to make a lot more money," Alexander explained. "We convinced distributors that this was smart business. And it was. When installing contractors came to people's houses, the customer would say, 'Holy Jesus, I can buy this Goodman unit for a lot less than I buy the Carrier unit.' And the business just exploded." In 1982, when Goodman began selling air-conditioning units, the company sold 50,000 units. By 2002, they were selling 1.2 million units a year. "We were a runaway train," Alexander said.

It probably didn't hurt that Goodman appreciated the product he was selling as much as anyone. "He was very fussy about his own comfort level," one former employee recalls. "But in the office, he didn't play with the thermostat himself. He left that up to his administrative assistant, whose name was Cynthia. I'd hear him shout, 'Cynthia, turn it up!' Or 'Cynthia, turn it down!' It was a little fetish of his."

I asked Alexander if there were any technological innovations behind Goodman's success. "The answer is no," he said bluntly. "It was a conventional air conditioner and everybody was making a conventional air conditioner and they still are. You know, you can get into some little idiosyncrasies to create a premium unit versus a basic unit and so forth, but they're basically the same today as they were fifty years ago, with a different refrigerant."

In Alexander's view, Goodman's brilliance can be summed up like this: "Don't ever kid yourself — price is everything."

At about the same time as Goodman was investigating the commercial possibilities of cheap cold air, writer William Faulkner died of a heart attack in Mississippi at age sixty-four. "The first fact of the day," wrote William Styron in an account of Faulkner's funeral in Oxford, Mississippi, "aside from that final fact of a death which has so diminished us, is the heat, and it is a heat which is like a small mean death itself, as if one were being smothered to extinction in a damp woolen overcoat." Styron described Oxford that day as a city drowned in "a heat so desolating to the body and spirit as to have the quality of a half-remembered dream, until one realizes that it has, indeed, been encountered before, in all those stories and novels of Faulkner through which this unholy weather — and other weather more benign — moves with almost untouchable reality."

Faulkner hated air-conditioning. He lived for most of his life in a two-story Greek Revival home built in 1848. Faulkner installed plumbing, electricity, and heat. But despite the monumental summer heat in Mississippi, he refused to install air-conditioning. Instead, he added a sleeping porch upstairs.* In his novel *The Reivers,* one of the characters grouses, "There are no seasons at all any more, with interiors artificially contrived at sixty degrees in summer and ninety degrees in winter, so that mossbacked recidivists like me must go outside in summer to escape cold and in winter to escape heat."

* The day after Faulkner died, his wife, Estelle, had a window-unit air conditioner installed in her upstairs bedroom.

Faulkner was not the only one who saw air-conditioning as a technology from hell. A cooling system was installed in the White House in the 1930s, but President Franklin Delano Roosevelt preferred to throw open the windows in the Oval Office during the summer and work in his shirtsleeves (in contrast, President Lyndon Johnson, who was a Texan, liked to crank up the air-conditioning and sleep under an electric blanket during sweltering DC summers). In 1945, Henry Miller titled his memoir about a trip across America *The Air-conditioned Nightmare.* Soul singer Aretha Franklin once stopped a live show when someone turned the air-conditioning on, worried that the cool air would ruin her voice.

But most people who tried it loved it. And it changed the landscape of America, opening up whole new frontiers to migration and development. White retirees in the North, previously unwilling to put up with the muggy heat of Florida and other Southern states, flocked south to air-conditioned condos on beaches and golf courses. They drove their air-conditioned cars to air-conditioned shopping malls and ate in air-conditioned restaurants. Corporations moved their headquarters south. Factories and manufacturing plants, fertilized by cheap real estate and non-union labor, blossomed in abandoned cotton fields.

The political ramifications of this demographic shift to the Sun Belt was enormous. The flood of conservative retirees to the South, once a Democratic stronghold, shifted the balance of power in American politics. Between 1940 and 1980, warm-climate states in the South gained twenty-nine electoral college votes, while the colder states of the Northeast and the Rust Belt lost thirty-one. Among the first to figure this out was Richard Nixon, who wooed these Sun Belt conservatives in the 1960s with anti–civil rights messages and racial dog whistles. American politics has never been the same since.

As the Sun Belt boomed, the technology of personal comfort had large and unexpected costs that were just beginning to be understood. In 1974, a group of scientists published research suggesting that chlorofluorocarbons (CFCs), the chemicals used as refrigerants in air conditioners, freezers, and refrigerators, as well as in aerosol spray cans, could deplete the Earth's ozone layer, which protected people (as well as plants and wildlife) from damaging effects of the sun, including skin cancer. In 1985, when a hole in the atmosphere was found over Antarctica, the ozone hole theory was no longer a theory. People freaked out. Within two years, an international treaty known as the Montreal Protocol was in place that cut the use of CFCs in half. The treaty was remarkably effective, a textbook example of the power of global agreements when everyone is on board. Today, CFCs are outlawed by 197 countries around the world and scientists agree that the ozone layer is slowly recovering.

Unfortunately, CFCs were replaced by another human-made type of chemical — hydrofluorocarbons (HFCs), which contain carbon, hydrogen, and fluorine. HFCs have the advantage of not destroying the ozone layer. But they have the disadvantage of being greenhouse gases that are up to fifteen thousand times more potent than CO_2. Air conditioners don't burn HFCs, but the gas often leaks out of the machines during repairs or disposal, or whenever the piping in the units becomes old and leaky. HFCs are being phased out over the next several decades, but the air-conditioning units that contain them will linger for a very long time.

Air-conditioning is also a big energy suck. Globally, air-conditioning accounts for nearly 20 percent of the total electricity used in buildings, which means it contributes a significant amount of the greenhouse gas pollution from buildings that are heating up the atmosphere. The hotter the planet gets, the more air-conditioning feels necessary. The more it feels necessary, the more electricity is required to power it. And as long as some portion of that electricity is generated by fossil fuels, that means more greenhouse gas pollution — which further heats up the climate.

It's a vicious cycle. And it is even more vicious in cities, especially in older and poorer neighborhoods, where old, inefficient window air conditioners hang out of every building, sucking heat out of the interior but blowing it out into the street. In this sense, air-conditioning is not a cooling technology at all — it's simply a tool for heat redistribution.

Harold Goodman died in 1995 at the age of sixty-eight. Some of his coworkers compared him to Sam Walton, the founder of Walmart, or Herb Kelleher, the founder of Southwest Airlines — populist businessmen who built empires by selling goods to the masses for the right price. "He was most proud of the jobs he created for people," his daughter Betsy told me. "The company he built put food on the table for thousands of people." By the time he died, Goodman Manufacturing was worth close to $1 billion.

For the next decade or so, the company was run by Goodman's son, John. A private equity firm eventually bought the company for $1.5 billion. They later sold it for $3.7 billion to Daikin Industries, a Japanese manufacturing giant that already had several lines of air conditioners and was looking for ways to expand its market share. The acquisition of Goodman made Daikin the largest air-conditioning manufacturer in the world. In 2017, Daikin consolidated all of its air-conditioning manufacturing, sales, and distribution on a five-hundred-acre campus about an hour east of Houston. The site is officially known as the Daikin Texas Technology Center, but it's sometimes referred to by a more poetic name: the Comfortplex.

The Comfortplex is one of the largest factories in the US (behind the Tesla factory in Austin and the Boeing Everett Factory in Washington). It covers ninety-four acres under one roof and employs seven thousand people. It's the Taj Mahal of air-conditioning and a monument not just to Harold Goodman's achievements, but to the ongoing human effort to control the Earth's climate one machine at a time.

Inside, the Comfortplex feels like a giant Costco, cheaply built to build machines that sell cheaply. Out on the production floor, fifteen-thousand-pound rolls of aluminum sheets are unfurled and stamped into louvers, heat exchangers, and other air conditioner parts. Robotic carts scoot around, carrying tools and parts. Newly built air conditioners roll down the production line, robotic soldiers ready for deployment against the enemy of heat. Seven production lines run 24/7.

Globally, the demand for air-conditioning remains insatiable. There are over one billion single-room air-conditioning units in the world right now — about one for every seven people on Earth. By 2050, there are likely to be more than 4.5 billion units, making them as common as cell phones today. Southern Europe, Indonesia, and the Middle East have all become addicted to cheap cold air. In Qatar, they even air-condition the outdoors: open-air stadiums built for the 2022 World Cup had cool air piped over the fields. In China, it was rare to feel mechanically chilled air twenty years ago. Now more than 75 percent of homes in Beijing and Shanghai have some form of air-conditioning. Over the last decade, 10 percent of the skyrocketing electricity growth in China has been due to cooling. In a country that is still largely dependent on coal for power, this is a climate disaster.*

As global air-conditioning dependence rises, the dangers of brownouts and blackouts rise along with it. During heat waves, when everyone cranks up the air-conditioning, the demand for electrical power spikes. "[In 2018] in Beijing, during a heat wave, fifty percent of the power capacity was going to air-conditioning," John Dulac, an analyst at the International Energy Agency, told *The Guardian*. "These are 'oh, shit' moments." Here in Texas, every heat wave is a nail-biter, with warnings coming from utilities to reduce power consumption or face rolling blackouts. On days when power demand is spiking, a small problem on the grid can easily cascade, threatening the stability of the

* Rising heat also has a big impact on AC demand. "If you want to cool your house to seventy-five degrees and the outside temperature increases from ninety-five degrees to ninety-eight degrees, that small change means you need one-point-three times more energy to cool," Andrew Dessler, a climate scientist at Texas A&M, told me. That's 30 percent more power, which means a 30 percent higher electricity bill, just for a three-degree increase in temperature.

entire system. And if power goes out for long on a hot day, businesses shut down, schools close, and people die.

Consider what happened in Hollywood, Florida, in 2017. A gentle sideswipe by Hurricane Irma knocked out power at a nursing home for several days, leaving it without air-conditioning. Outside, the temperature was only in the mideighties — hardly a heat apocalypse. But inside the nursing home, in the poorly built, poorly ventilated, air-conditioning-dependent building, the temperature soared, especially on the upper floors. The nursing staff ignored the slowly broiling patients. It wasn't until two days after the power failed that someone finally called 911. When Lieutenant Jeff Devlin from the Hollywood Police Department arrived, "It was markedly hotter in the inside than the outside," he later testified in court. "The smell of urine and feces immediately hit me." Twelve patients died, some with body temperatures as high as 108 degrees.

Air-conditioning is a distinctly American invention, as red, white, and blue as a double cheeseburger with a Coke and a side of fries. And like hamburgers and Cokes, it quickly went from an American curiosity to a global addiction. "Comfort is valued because it promises consistency, normalcy, and predictability, which allow for increased productivity or a good night's sleep," architectural historian Daniel Barber wrote in an essay about our addiction to air-conditioning. "Comfort indicates that one has risen above the inconsistencies of the natural world and triumphed, not only over nature and the weather, but over chance itself. We can rely on comfort. It will be there when we get back."

It's a false victory, however. The quest for comfort at all costs — or to be more precise, the sense that comfort is an inalienable right of modern life — is wreaking havoc on our world. As Barber put it, "Comfort is destroying the future, one click at a time."

There are ways to limit the damage. The most obvious one, which I've mentioned earlier and will say again: stop burning fossil fuels and move to clean energy. That may happen in some places faster than you think (at least for electricity generation). But it will also happen in some places more slowly than you might hope. So increasing the efficiency of air conditioners can help (in the US, new efficiency standards take effect in 2023).

Another way is to think differently about how we build things. The rise of air-conditioning accelerated the construction of sealed boxes, where the building's only airflow is through the filtered ducts of the air-conditioning unit. It doesn't have to be this way. Look at any old building in a hot climate, whether it's in Sicily or Marrakesh or Tehran. Architects understood the importance of shade, airflow, insulation, light colors. They oriented buildings to capture cool breezes and deflect the worst heat of the afternoon. They built with thick walls and white roofs and transoms over doors to encourage airflow. Anyone who has ever spent a few minutes in an adobe in Tucson, or walked on the narrow streets of old Seville, knows how well these construction methods work. But all this wisdom about how to deal with heat, accumulated over centuries of practical experience, is all too often ignored. In this sense, air-conditioning is not just a technology of personal comfort; it is also a technology of forgetting.

In the end, the most enduring legacy of air-conditioning may be the divide it has created between the cool and the damned. And the hotter it gets, the bigger that gap will grow. This is not a technological failure as much as it is a cultural and psychological issue. The simple truth is that in the second half of the twentieth century, prosperous Americans got hooked on comfort, with little thought about the cost of that comfort to others, to the welfare of other species, or to the world around them. That addiction has now spread to millions of people around the world, who find they too cannot live without cheap cold air.

12. What You Can't See Won't Hurt You

In June of 2021, *The Telegraph,* a UK-based newspaper, assigned thirty-three-year-old Pakistani photographer Saiyna Bashir to accompany reporter Ben Farmer to Jacobabad, a city of two hundred thousand people in Sindh province in central Pakistan, for a story about extreme heat. It was a story she had been wanting to do for several years, in part because as a photojournalist she was interested in chronicling the lives of people whom the rest of us often choose not to see: families living with HIV, Pakistani women scarred from acid attacks, children wandering alone in camps for Afghan refugees. People living in extreme heat fascinated her in a similar way.

If there was anywhere in the world you would go to photograph heat, it would be Jacobabad. By any metric, it is one of the hottest cities on Earth. A few weeks before Bashir's visit, the temperature had hit 126 degrees every day for more than a week. Worse, it was a thick, wet heat — the deadliest kind. And there was not much relief. Two hundred and twenty million people live in Pakistan, but there are fewer than a million air conditioners in the country.

It's important to underscore that killer heat is not something Pakistanis brought on themselves. Pakistan produces about one half of one percent of the world's CO_2 emissions. On a per capita basis, each Pakistani is responsible for less than one-fifteenth as much CO_2 as each American. It's how the climate crisis works: the rich pollute, the rest suffer.

Bashir was born in Karachi, but because her father was in the military, he was constantly getting transferred — "I grew up all over the country," she told me. She began taking pictures when she was a teenager but never really dreamed she could make a career out of it. In 2014, she left Pakistan to study journalism at Columbia College Chicago and ended up getting a job as a photographer at a newspaper in Madison, Wisconsin. She donned a gas mask and photographed the riots in Ferguson, Missouri, that erupted after the fatal shooting of Michael Brown by police officer Darren Wilson. She got screamed at by Trump supporters while photographing his rallies during the 2016 campaign. She photographed homeless people freezing on the streets of Chicago. Her photos won awards and gave her the confidence in her abilities. She moved back to Islamabad, got married, and now works as a freelance assignment photographer for the *New York Times,* the *Washington Post,* and many others. She sees herself following in the footsteps of great women photojournalists like Carol Guzy and Lynsey Addario who cover war and humanitarian crisis around the world.

For a photojournalist, heat is a difficult subject. What is the visual reference for heat? A glaring sun? A melting ice cube? The story that needs to be told about heat is the story of what it does to life, human and otherwise. But how do you make visible the story of an invisible killer?

Like most journalists, Bashir has a fundamental interest in chronicling what extreme conditions reveal about human character. Her work is a way of expressing her awe at the courage and toughness of people. But her work is important for other reasons too. A single great picture can change the world. The Blue Marble photo taken from Apollo 17 in 1972 gave millions of people a new perspective on our place in the universe and helped inspire the environmental movement. John Filo's photo of the woman kneeling over the body of a Kent State University student after he was shot by the National Guard during a protest against the invasion of Cambodia in 1970 changed how millions of Americans thought about their government. Once you have looked at "Falling Man," the photo by Associated Press photographer Richard Drew of a man plunging headfirst off the World Trade Center, you'll never think about the September 11 attack the same way again. Pictures of Martin Luther King Jr.'s Selma-to-Montgomery march remain touchstones in the fight for justice and equity. And I don't mean that in a sentimental way. These images have changed laws, shifted politics, redefined our past and our future.

But there are no iconic images of extreme heat. The difficulty of photographing heat is obviously one reason for that. Another may be that our cultural awareness of the risks of extreme heat is so rudimentary that it's not seen as a worthy subject. To photograph heat, you have to see it in a way that goes beyond sweat or melting ice. Photojournalists like Bashir do not pretend to be artists, but sometimes their work can approach art — that is, they can create a photo that transcends the surface reality of the moment and captures something deeper about human nature and human suffering.

Bashir and Farmer and their driver headed out of Islamabad at 7:30 a.m. in a rented Toyota that, blessedly, had air-conditioning. It was an eight-hour trip north along the floodplains of the Indus River. It was a better road than it used to be, now that the Chinese had poured money into Pakistan for road development and improvement. After driving all day, they stayed in a small hotel near the city of Larkana. The next morning, they were up early, traveling through a parched, treeless landscape of cotton and rice fields. The road was crowded with shepherds tending their goats, trucks overloaded with old TVs and furniture, motorbikes buzzing by like swarms of flies. They stopped so Bashir could take photos of a man under a red-and-yellow awning who was selling *thadal,* a drink made of water, sugar, milk, dried fruits, pepper, and almonds that many Pakistanis (including Bashir) believe help with cooling. He kept the *thadal* in an orange cooler, ladling it out in plastic cups to sweaty motorbikers who stopped to enjoy some shade. It was only 9 a.m., but the temperature was already more than 100 degrees.

It was forecast to hit 115 later in the day — hot, but not record-breaking hot.

A half hour later, they entered Jacobabad. It was a tangled nest of heat and commerce: sketchy-looking banks, fruit stands, a drugstore, a spaghetti of wires overhead, Sindhi pop blaring from donkey carts, the smells of cardamom and motorbike exhaust. Victoria Tower, a monument to colonialism that was built in 1887, stood like a dried-out skeleton in the center of the city. Everyone was dressed for heat: both men and women wore colorful *kameez* (a type of long tunic or shirt) made of cotton lawn, and loose, billowy pants. Women wore a *dupatta,* a traditional scarf, over their heads.

There was no shade, no relief. Whatever trees had once grown in the city had long ago been cut down for firewood — what was left was just a few saplings here and there that someone had planted with the vague hope that they could escape the axe and someday provide relief from the sun. Only local officials, the police, and the hospital are lucky enough (and wealthy enough) to have air-conditioning. But even for them, the power is so unreliable that they can't depend on it. Not long ago in Sahiwal, a neighboring city, the extreme heat combined with a power outage had killed eight babies in a hospital ICU when the air-conditioning cut out.

The night before, Bashir had made a list of things she wanted to look for and photograph. She knew, for example, that most Pakistani towns have a local ice factory. That might be promising. But mostly she just needed to keep her eyes open. Like most good journalists, she doesn't know what she is looking for until she finds it.

Bashir and Farmer headed first to the town square, where men were selling ice from wooden carts shaded by plastic tarps propped up with bamboo poles. Each cart had five or six blocks of ice, pushed close together to slow the melting. People lined up to buy small chunks of ice to take home to use in their water coolers so their families would have cold water to drink during the worst of the heat. In exchange for a few rupees, the ice men used mean-looking machetes to hack off shards of ice, which they put into a used plastic shopping bag for people to carry home. Customers smiled — having a bag of ice felt lucky, even hopeful. Some reached in with a dirty hand and grabbed a small piece and slipped it into their mouth. Nearby, donkeys stood quietly, staring at the ice, their ears pinned back as if the heat were pressing down on them. Bashir snapped away with her Canon EOS 5D — a piece of technology that, she knew very well, was worth more than any of these men earned in a year. She didn't feel bad about that. But it did inspire her to take her job seriously. It was a privilege to be on the other side of the camera.

As midday approached, the temperature hit 110 degrees. Bashir, following Muslim tradition, had to keep herself covered: she wore a black-and-white *kameez,* a white *dupatta,* and black pants. An hour or so was all she could handle before she had to take refuge in the air-conditioned Toyota.

They drove out to the ice factory, which was in what looked like a dilapidated warehouse in an industrial section of the city. Bashir and Farmer stepped into a big, open room with a pulley on the ceiling and steel doors on the floor. It was blessedly cool in there. One of the workers explained to Bashir the process of ice-making — it's just like a giant refrigerator, compressing fluid to heat it up, then letting it expand and chill. He showed her the big radiator tubes outside, where the heat is released, then the noisy room where the compressor runs, a big wheel spinning. She photographed it all — the loud, clanking, Dickensian machinery of cool. She shot workers rolling the pulley over the steel doors in the floor and using the hoist to haul five-foot-long pillars of ice out of the floor. The best images were not of the ice itself, but of the fatigue on the faces of workers who were sitting against the wall, worn out from manufacturing ice in a hot, hot world. She glanced at the screen of her Canon. The images were good. But were they good enough?

Bashir and Farmer moved on. In an informal settlement nearby, Bashir saw a woman lying on a woven bed called a charpoy. Charpoys are well engineered for heat, with the jute webbing elevated a foot or so off the ground to allow air to circulate beneath it. Bashir asked the woman, whose name was Shama Ajay, if she could photograph her. The woman nodded but didn't move. She lay on her side, staring at Bashir's lens. She was young, maybe midtwenties, wearing a wine-red *kameez* with beautiful embroidery on the front. She looked almost liquid in the heat, one hand resting on her stomach, the other near her head, her dark-brown eyes looking straight at the camera, as if to suggest that you — whoever you are, looking at her photo — are complicit in all this.

As Bashir and Farmer traveled through the city, Bashir found moments worth capturing: old men drinking tea at an outdoor café. A boy selling handmade jute fans on the street. A man sprinkling water on tomatoes and okra and potatoes at a vegetable stand to keep them cool. Barefoot workers at a rice mill, raking a fifteen-foot-high pile of golden rice and loading it into sacks. And in the market, a man dressed in white selling extension cords and small rotating fans, which he ran off a portable battery on the ground beside him. They were all good images, capturing life in the city of heat. Many of them showed human resilience and ingenuity and strength. Still, they were not quite good enough.

Finally, their guide suggested they go to the water station, where men (and in a rigidly patriarchal society like Pakistan, they are all men) fill donkey carts with blue plastic containers of water, which they then sell for fifty rupees (about twenty-five cents) to people in their neighborhoods.

The water station was on the outskirts of the city and amounted to little more than a concrete wall with five water pipes attached to it at shoulder height. On each pipe was a rubber hose. Water vendors load up their carts full of containers of water, which they then delivered around town. When Bashir arrived, there were several donkey carts at the station, and she photographed them going about their work. The images weren't very interesting.

Then she watched a man leave his donkey cart to the side and walk up to the plastic hose alone. He was middle-aged, with a beard and faraway eyes, wearing a plain ivory-colored *kameez* and bright orange-and-blue rubber sandals. His name, Bashir would later learn, was Mehboob Ali. Ali did something very unusual for a Pakistani man to do in public, even in the extreme heat of a summer afternoon: he sat down and grabbed the water hose and raised it over his head and soaked himself with water. It was as if he were sitting under a waterfall — he was completely soaked, sitting there on the concrete, everyone looking at him. Bashir started snapping pictures. He didn't even notice her. He moved the hose around above his head, water splashing over his face.

In one image, a close-up of Ali's face, his hands raised above him holding the hose, his eyes closed, she got something special. Water hangs like icicles off his moustache and beard. His face, covered with water, reveals a bliss that is the inverse of the pain he endured from the heat. It feels like a very private moment, the transition between a man's suffering and his relief. The photo captures the brutality of heat while also suggesting redemption from it. "I like this image because of the calmness in his expression," Bashir told me later, after I had seen the photo in *The Telegraph* online. Even in one of the hottest and most hellish places on Earth, Bashir's photo suggests, you can be saved.

The way we communicate about extreme heat is often distorted by nostalgia for a climate that no longer exists. In July 2022, when the high temperature in Austin shattered records that went back to 1898 by eight degrees, the local news station illustrated the story of the heat wave with images of people playing with dogs in the park. It reminded me of a *New Yorker* cartoon I'd seen of two people standing below a fire-breathing dragon that is torching their house: "I know that it comes earlier every year, and is destroying the future for our grandchildren, but I'll be damned if that extra bit of warmth isn't nice."

Part of this distortion has to do with the simple fact that people love warm weather. I've spent a lot of time in very cold places, from Greenland to Antarctica, and I find them exotic and thrilling. But how many ads have you seen from airlines and travel companies of hammocks and lounge chairs covered with ice? We are constantly bombarded by images that suggest that if paradise does exist, it is warm and sunny.

And in some sense, that's not surprising. Love of heat is in our genes. A recent study compared temperature and air moisture levels in thirty-seven US homes to outdoor climates around the world. They found that in all but three of the homes, the preferred temperature was 72 degrees, with low humidity, a combination that most closely resembled the temperature and humidity in East Africa — the same region of the continent where the first humans lived hundreds of thousands of years ago. As Mark Maslin, a paleoclimatologist at University College London, observed, the findings suggest that even when people can set the temperature and humidity at whatever they want, "they then choose something that harks back a hundred thousand years to Africa."

Adding to the complexity of how we communicate about heat waves are the various measurements and indexes that are used to quantify heat. There is temperature and humidity, of course. But there is also a heat index, "feels like" temperature, "apparent temperature," wet bulb temperature, and wet bulb globe temperature, as well as proprietary indexes such as RealFeel that are pushed by companies like AccuWeather (see the glossary for definitions of these various heat metrics). It doesn't help that most of the world uses the Celsius scale for temperature, while the US (along with the Cayman Islands and Liberia) uses the Fahrenheit scale. All in all, it's no surprise that people are confused about when heat crosses the boundary from something that feels good to something that can kill you.

One big problem with communicating about heat waves is simply defining what a heat wave actually *is*. A tropical storm or a hurricane is defined by wind speed. A drought is defined by lack of rain. But what defines a heat wave? Is it 105 degrees? Is it 110 degrees? And how long does it have to last — an hour? A day? Three days? And what about humidity — how do you factor that into the definition of a heat wave?

Here again, the invisibility of heat contributes to the problem. Visually, we all know what the spinning eye of a hurricane looks like, and how as the wind speed of the storm increases, the eye of the storm expands. For weathercasters, it is fairly easy to project the path of the storm as it strengthens. But heat waves have no visuals. There is no spinning eye, no trajectory. Weathercasters talk about things like "heat domes," but that's a metaphorical description, based on the buildup of high pressure in a certain area — you can't track the movement of a heat dome, and you can't look out your window and see it.

With hurricanes, the question of which storms to name is purely meteorological: if the wind speed rises above thirty-eight miles per hour, the storm gets a name. It is simple and straightforward.

But there is no comparable metric for heat waves. It is not just that a 99-degree day in Buffalo feels very different from a 99-degree day in Las Vegas. It's that the people who live in Buffalo are less likely to have air-conditioning, and so they are more vulnerable. People in Buffalo are likely to know less about how to deal with heat than people in Las Vegas and are less likely to check on family and friends who might need help. In short, heat waves are more like stories than meteorological events. Each one has a particular setting, a cast of characters, and different dramatic flash points.

In this sense, heat waves are the opposite of earthquakes. For centuries, earthquake intensity was measured by how much the chandeliers swung or how many houses fell down. Then in 1935, American seismologist Charles Richter developed the Richter magnitude scale to describe the intensity of an earthquake. Richter used a seismograph to measure actual Earth motion (seismic waves) during earthquakes, then plotted these waves on a logarithmic scale in which each point was ten times greater than the previous (a magnitude 7 is ten times greater than a 6, and a hundred times greater than a magnitude 5). It was the first ranking of a natural event that measured risk in a scientifically rigorous way.

In contrast, the heat wave ranking system used by the National Weather Service (NWS) has very little scientific rigor. The NWS ranks heat in three categories: watches, warnings, and advisories. Heat watches are the least severe in this ranking system; heat advisories are the most severe. But exactly what defines each of those categories is left up to local NWS field offices. For example, this is the guidance that NWS offers local field offices for an Excessive Heat Warning:

> Heat Index values forecast to meet or exceed locally defined warning criteria for at least two days (Typical values: 1) Maximum daytime Heat Index ≥ 105°F north to 110°F south and 2) Minimum nighttime lows ≥75°F).

Many things are confusing about this system. Who can keep straight the difference between a warning, an advisory, and a watch? I can't, and none of my friends and family can either. Why is a heat index of 105 degrees — as opposed to, say, 104 or 106 — the trigger point for a heat advisory in the North? And where exactly do they draw the line between North and South? The NWS says the vagueness is deliberate, so local conditions can be factored into the rankings. "We leave a great deal of discretion to local field offices," Kimberly McMahon, a NWS Public Weather Services program manager, told me. But the end result is too often confusion and a lack of urgency.

Another problem: there's not much evidence that these warnings make much difference. One 2018 study found that NWS heat alerts were associated with a statistically significant reduction in mortality in only one of the twenty cities studied. Moreover, there is no standard for heat alerts, and each of the 116 NWS field offices is free to make its own decisions about what metric to use for alerts and when to issue them, leading to some places getting lots of heat alerts and some places getting very few. Making matters worse, the alerts are not issued in places that need it most. As the study put it: "The spatial pattern in heat alerts was not correlated with heat attributable mortality, suggesting that the current approach may not well align with heat-health risk."

Nobody should die in a heat wave. People die because they are alone and don't know what to do and don't ask for help. Or they don't have air-conditioning (or the money to run it). Or they can't get to a cooling center. Or they are afraid that their employer will fire them if they stop working.

Moreover, people die because they don't understand the warning signs of heat exhaustion and heatstroke or don't take precautions and ask for help when those warning signs appear. There is so much ignorance and confusion about what to do in extreme heat situations. Do you turn on a fan and open the windows? How much water should you drink? Should you take a cold bath? Is it good or bad to sweat a lot? If your heart starts racing does it mean you're going to have a heart attack?

This is the basic problem that Kathy Baughman McLeod understood when she decided to make extreme heat the focus of the Adrienne Arsht-Rockefeller Foundation Resilience

Center.* Arsht-Rock, which was founded in 2019, is a nonprofit venture that started with only one directive: to increase the resilience of people in a world ravaged by climate change. Baughman McLeod could have focused her attention on a number of deserving issues, from food security to hurricane preparedness. But after thinking about it for a few months and talking with other climate groups, she decided that the risks and impacts of extreme heat were not getting anywhere near the attention they deserved. In fact, those risks and impacts were practically invisible. Making them visible, Baughman McLeod thought, could save millions of lives.

Baughman McLeod, fifty-three, is a dynamo. She has hazel eyes and a quick laugh and a way of speaking that can shift quickly between lighthearted banter and drop-dead seriousness. "I want to innovate, evolve, and improve the world at the same time," she told a journalist early in her career. "I want to make a difference."

From the beginning, Baughman McLeod understood two important things: the climate

* Full disclosure: I'm a senior fellow at Arsht-Rock. The fellowship is unpaid, but I do sometimes indulge in a free glass of wine at Arsht-Rock's annual meetings.

crisis was happening fast, and it was going to change our economy in a big way. She spent a decade or so working in various climate- and finance-related jobs in and out of Florida state government, including a stint at the Florida Energy and Climate Commission. In 2013, she became the managing director for global climate resilience at the Nature Conservancy, then moved on to become a senior vice-president for environmental and social risk at Bank of America. At the Nature Conservancy, Baughman McLeod was known for innovative ideas, like an insurance policy she created for the coral reef in Quintana Roo, Mexico, that protects the reef and beaches in the region — as well as the $10 billion in annual tourism revenue they generate — from damaging storm surges. It was the first natural structure in the world with its own insurance policy — the *New York Times* called it "a radical experiment in finance."

At first glance, it's hard to see how innovative financial tools work with heat. It's not like solar panels, where an entrepreneur can get a return for investing in new technology. What entrepreneur is going to invest in better bus stops? There is no money to be made in planting urban trees for shade or opening cooling centers to give vulnerable people refuge during a heat wave. To put it another way, if extreme heat is the enemy, what does your fighting force look like?

One answer came to Baughman McLeod in a nondescript meeting room in Sacramento in the fall of 2019. She was there for a gathering of a working group in California on climate and insurance, where a dozen or so people met to make recommendations to California's insurance commissioner, Ricardo Lara. Baughman McLeod was chatting with Kristen Torres Pawling, who is the sustainability program director in Los Angeles County. Pawling had been struck by what a hot summer it had been, and the wildfires that had burned nearly three hundred thousand acres of the state. Pawling commented on how everyone knew about the fires and talked about them by name, but heat waves, which killed far more people than wildfires, were much harder to discuss.

"Why don't we name heat waves like we name storms and wildfires?" she asked Baughman McLeod, not really expecting an answer.

And that's when a lightbulb went on in Baughman McLeod's head.

Naming is a basic human impulse. We name our kids, our pets, our cars, our houses, the mountains we climb, and the constellations in the night sky. Storms and hurricanes have been named for centuries. Sometimes, they were named after prominent victims, like Solano's Storm, which wiped out the Spanish fleet off the coast of Florida in 1780, and was named after the Spanish commander of the fleet, José Solano y Bote (he survived, but thousands of men under his command didn't). Often storms were named after the year they struck and the area they hit, such as the "Great Miami Hurricane of 1926."

In the 1950s, in a nod to the maritime tradition of naming ships after women, the NWS started naming hurricanes after women. It had nothing to do with increasing public awareness of how destructive hurricanes can be. It was mostly to improve communication among ships, airplanes, and meteorological stations, which in the past had used mainly longitude and latitude to define storms.

But naming storms after women was not such a bright idea. In the 1960s, women were not happy with the gendered and misogynistic way in which female-named hurricanes were described as "witches," "capricious," "furious," and "treacherous." (One outspoken feminist recommended that hurricanes be called "himicanes" instead.)

And so it was that from 1979 onward,

Atlantic tropical storms and hurricanes have alternated between male and female names. Today, the naming of hurricanes is handled by the World Meteorological Organization (WMO), a United Nations agency that oversees weather monitoring and forecasting.

But where and when and how do you name a heat wave? Of course, people had been doing it on their own for years. As an editorial in a Jackson, Mississippi, newspaper quipped: "In Mississippi and the rest of the Deep South, heat waves already have two names: July and August." When a heat wave boiled Italy in 2017, Italians nicknamed it Lucifer. That inspired a writer for the *Palm Beach Post* to point out that "Heat Wave" is an actual character in the DC Comics pantheon of supervillains. "We could continue that theme by calling our long summer Blisterer, Scorcher, Sizzler or Steam Bath, who would be a blundering sort of evil-doer."

Baughman McLeod thought there was a better way to do it. In August of 2020, in the middle of another hot summer, she launched a group called the Extreme Heat Resilience Alliance with thirty global partners, including the Red Cross and the cities of Miami, Florida, and Athens, Greece. "This extreme heat crisis can no longer be the 'silent killer' it is," Baughman McLeod said at the launch.

The alliance would help build a movement to take on heat. Later, with those partners and others, Arsht-Rock created new positions for chief heat officers in cities, and a heat health science panel, and developed a tool kit for policy makers to better understand what they can do to reduce heat mortality and morbidity. But Arsht-Rock's top priority was ranking and naming heat waves. "Naming heat waves is the clearest way to communicate the dangers and severity of this risk, which is growing," Baughman McLeod told the *Washington Post.*

Not surprisingly, Baughman McLeod's naming and ranking proposal generated pushback from academics and traditionally minded scientists who did not appreciate her bold ambition and outside-the-box thinking. Forty-two heat researchers signed a letter to Baughman McLeod warning her that naming and ranking heat waves "does not align with published global heat-health priorities and may be distracting, or even counterproductive."

In some sense, the opposition from the science establishment was not surprising. Because naming a heat wave is *not* science. "It is branding," Baughman McLeod says unabashedly. "It is PR. And it's PR that will save lives."

■■■■

But it's *also* science, because before you can name a heat wave, you must predict which one will be dangerous enough to merit a name. Laurence Kalkstein, a respected heat researcher working with Arsht-Rock, recommended a system for ranking heat waves that uses much more than meteorology. It is built around the expected health impacts of the heat to a specific community. Kalkstein had developed a system called spatial synoptic classification, which identifies eight different types of air masses: dry tropical, moist tropical, etc. He looks at cities and sees how these different air masses correlate with data from all causes of mortality for the region. He can see that in, say, Albuquerque, when a dry tropical air mass moves in, the mortality rate spikes by 15 percent. If he does this correlation enough times with air masses in any given city, he can get a pretty good estimate for how many people that air mass will kill whenever it arrives.

"We can forecast the air masses up to five days in advance using basic meteorological data," Kalkstein told me. "And then we can develop specific algorithms to see how many people are going to die."

Ranking systems are always imperfect.

But Kalkstein's system has two advantages. First, it looks exclusively at mortality, rather than other indicators of health impacts, such as emergency room visits. Second, it blends several factors, including humidity and nighttime temperatures, which have a big effect on heat-related mortality, into a single score. Third, it is based on actual past history in an actual place, so it is working with data that is very specific to a city or region. All Kalkstein and his team needed was past weather and mortality data, and it could all be crunched together in an algorithm that would calculate future mortality with the forecast of any given air mass.

Deciding which heat wave to name is one thing. Deciding what to name it is something else. Baughman McLeod hired a social research firm to conduct focus groups in several regions of the country to test different naming schemes. Among them: Greek gods (Zeus, Apollo, Hades); Greek letters (alpha, beta, gamma, delta); hot foods (salsa, chili, pepper); locations (Camelback Mountain, Lookout Point); color (white, orange, red); cooking (rare, medium, well done).

In the focus group that I observed, which was made up of Arizona residents, there was no broad agreement on the best naming scheme. In fact, the participants didn't even think naming was a good idea until they were told that heat waves are the deadliest extreme weather events, killing far more people every year than hurricanes or floods. Then they agreed, yeah, naming might be important. As for what to name them, Greek mythology "is evocative," one man said, "but it's also kind of demoralizing. It makes heat sound like a supernatural force." A woman added: "Using mythological names makes you think we don't have control of what's happening, but we do."

Cities were the ideal labs for a new heat wave ranking and naming system, in part because both the politics and the science were more manageable than trying to roll the program out on a state or nationwide scale. Still, there were a lot of moving parts. In the spring of 2022, nearly a year after Baughman McLeod launched Arsht-Rock's heat naming and ranking initiative, Athens city officials decided to rank heat waves using Kalkstein's air mass methodology, but skip the actual naming and just use color-coded alerts.

In Seville, on the other hand, city officials committed to a pilot program to both rank and name heat waves. And for good reason. The city of nearly seven hundred thousand people in southern Spain is regularly baked by extreme heat. And it's getting worse. In recent decades, the frequency of heat waves in Spain has doubled. To help sort through the political and scientific complexity of ranking and naming, Baughman McLeod and her colleagues joined forces with the city to put together a group called the proMETEO Seville project, which was an alliance with universities, the Seville mayor and city council, and AEMET, the Spanish Meteorological Agency.

The various agencies agreed that the pilot program would use Kalkstein's algorithm to categorize heat waves based on the potential impact on human health and mortality. There would be three categories, and only the deadliest, a category three heat wave, would get a name. There was discussion about exactly where to draw the line for a category three heat wave: A 30 percent increase in mortality? A 45 percent increase in mortality? Also, there was debate about whether that projected mortality number should be disclosed in the heat wave ranking. That's not a surprise. Does any politician want to tell his or her constituents that some of them are going to die over the next few days? After much discussion, city officials chose a simple naming scheme that alternated between male and female names in reverse alphabetical order: Zoe, Yago, Xenia, Wenceslao, Vega, etc.

The initiative turned out to be well timed. In 2022, Spain sweltered through one of its earliest heat waves on record (May ranked as the hottest in fifty-eight years). In June, when the temperature regularly hit 107 degrees, the heat coincided with hatching season for swifts, a bird that often builds its nest in building façades or the cavities of roofs. "Our buildings are usually made of concrete or metal plates and these get very hot," said Spanish biologist Elena Moreno Portillo. "So it becomes an oven and the chicks, who can't fly yet, rush out because they can't stand the temperature inside. They're literally being cooked." In Seville, Portillo said, "You would walk down the street and there would be a hundred chicks, lying at the foot of a building, some dying and some barely alive."

In mid-July, the heat started to rise — not just in Seville, but throughout southern Europe. "The average number of deaths in Seville is about between fourteen and fifteen a day," Kalkstein told me later. "During the heat wave, we had many days when deaths numbered in the twenties. And some that were in the thirties."

On July 24, the forecast showed that the heat in Seville was about to spike. Temperatures were forecast to soar above 109 degrees, with high nighttime temperatures. On Kalkstein's ranking system, this was still only a category two event. But for officials in Seville, it was close enough. They made an announcement at 9 a.m.: Heat Wave Zoe had arrived.

The name was important. But what was really important was a series of alerts, warnings, and other messages on social media to let people know how they could protect themselves from the heat, including simple, clear instructions about reducing physical activity, staying inside, closing the blinds during the day and opening windows at night to ventilate, drinking fluids, and avoiding heavy meals.

Heat Wave Zoe sparked intense media interest both in Spain and internationally. If the strategy was to raise the awareness of heat and educate people about how to handle it, then by that measure, the project was a resounding success.

Did naming the heat wave Zoe actually save lives? "We are at early stage," said Jose Maria Martín Olalla, an associate professor of physics at Seville University who worked on the project. "We applied the name only in the city of Seville, so it was a very local thing. I think the question is, how well will it work in the long term? We need to alert the population of the dangers we are facing, and which will get worse in the future. And I'm convinced that a very good way to do that is by giving names to heat waves."

In fact, a survey Arsht-Rock commissioned a few months after Zoe of more than two thousand residents in seven regions of Spain found that people who recalled hearing about Zoe were more likely to engage in safe behaviors like drinking water and working from home to avoid the heat. They were also more likely to talk about it with others, and to believe that government was working to protect them. The link between raising awareness and reducing risk has been underscored by another study that looked at what happened after US embassies started tweeting air pollution statistics in the cities where they were located. Researchers found that this very cheap and easy investment led to substantial reductions in air pollution and better health.

To Baughman McLeod, the Seville pilot project was an important first step that helped build evidence and momentum for others to think more deeply about how they communicate about heat. Before the summer was over, a half dozen other cities were in the process of setting up a heat-health warning program based on Arsht-Rock's ranking and naming system. The WMO issued a technical brief examining the pros and cons of naming heat waves, concluding that the organization "should initially look to conduct an evaluation of the effectiveness, benefits, challenges and sustainability of existing initiatives to name heat waves, using the findings to inform any future proposals." In Sacramento, where it all began for Baughman McLeod, the legislature passed a bill that directs the California Environmental Protection Agency to develop a health-based ranking system for heat waves. The new law meant that forty million people would now have a better idea of how dangerous future heat waves would be, and what actions they should take to protect themselves.

Baughman McLeod is also returning to her roots in finance. Working with her team at Arsht-Rock and an India-based group called SEWA (Self-Employed Women's Association), Baughman McLeod is testing an innovative micro insurance program that compensates women in the Global South for lost income during extreme heat events. The program, which pays women in advance of dangerous heat waves, combines an early-warning mechanism, training, and cash payouts, based on forecasted heat conditions. It also includes supportive, practical interventions like water, electrolyte tablets, and gloves. The goal is to give women the financial security to avoid dangerous work during a heat wave and take care of themselves and their families.

"We think that the conditions and the death tolls push us to accelerate what we're doing to save lives as soon as we can," Baughman McLeod told me. "I mean, people are dying out there. We need to do something."

13. Roast, Flee, or Act

Once upon a time, summers in Paris were a bit like summers in Seattle. The temperature drifted around in the seventies. It rained every so often. The humidity wasn't too bad. The weather was like that for centuries — which is why, until recently, virtually no one in the city had air-conditioning. Why bother? Besides, there's a long tradition in France of taking August — usually the hottest month of summer — off for holiday. The city virtually shuts down and people go to the beach in Brittany or the mountains in the Alps to cool off and relax. Think of it as an old-fashioned adaptation to heat.

People who stick around during August are often older, or have jobs that require them to stay and keep the city functioning for the thousands of visitors who want a selfie at the Eiffel Tower. In the summer of 2003, Parisians who remained in the city were hit with something they had never considered before: a heat wave. There had been hot days in Paris, but nothing like this. For nine days in August, the daytime temperature was above 95 degrees, sometimes spiking up to 104 degrees. It didn't cool off much at night either.

It took a few days for the full scope of the tragedy to reveal itself. Police and fire officials began responding to more and more calls. Hospital emergency rooms started to fill up. A week or so into the heat wave, city officials began running out of places to store bodies. The health ministry wanted to put them on public ice rinks near the capital, but the rinks were closed in August and it would take too long to refreeze them. So authorities erected refrigerated tents in parks. But there were still too many bodies. Even working around the clock, burials and cremations could not keep up with the number of deaths. The city extended the maximum delay between death and burial from six to fifteen days, but that only resulted in what one report called "a massive engorgement of sites for the accommodation of bodies." Finally, city officials commandeered a food storage warehouse. They also leased or purchased refrigerated food trucks. "One truck," a writer recalled, "stripped of its decals, still bore the silhouette of the name of the butcher from whom the city purchased it."

In less than two weeks in 2003, fifteen thousand people in France died as a direct result of the heat wave. Nearly a thousand lived in central Paris. Many of the victims lived alone, in top-floor garrets, or attic apartments, where the heat built up beneath zinc roofs and literally cooked people as if they were in an oven. It took weeks for all the bodies to be recovered. Entire apartment buildings had to be evacuated because of the pervasive smell of death.

Many Parisians who had been out of town returned to gruesome scenes. One twenty-year-old woman had been warned before she returned that a neighbor had died in the apartment above hers. But when she opened her front door, she screamed. On the floor was "a pool of dried blood, blood from a body, everything . . . urine, blood, everything." The body, it turned out, had been in the apartment above hers for more than a week before it was discovered. Bodily fluids had trickled down the walls and through the slats in the paneling on the ceiling. Vases in her kitchen, she discovered to her horror, were filled with liquid.

"I had to vomit," the woman said. "I was nauseated. I spent practically the whole afternoon under the shower washing my-self. In the apartment, the smell was still so strong, even months later, it never went away. It was impregnated in the couches, in the bed, everywhere. For several months, every time I went there, I retched."

Twelve years later, in December of 2015, I was in Paris to cover COP21, the United Nations climate summit. I stayed in a garret apartment in the Fifth Arrondissement that I'd rented on Airbnb. It was a small, cozy place on the sixth floor with a low ceiling and big wooden beams. It felt medieval, even though I knew it wasn't. When I looked out the window, I saw an ocean of Paris's famous zinc roofs on top of the eighteenth-century buildings that surrounded me. I thought about how lovely they looked, especially in the evening. The *New Yorker* writer Alexandra Schwartz captures it perfectly: "In the blue hour of middle evening, just after the sun has set but before the light has finished draining from the streets, the roofs glow blue, sometimes so intensely that the blank walls below pick up the color and reflect it, giving the city a submerged quality, as if it had sunk quietly to the ocean floor."

At that time, I had never heard of the 2003 heat wave, even though I had been covering the climate crisis for more than a decade. I had no idea that it was in rooms like this, high in the old buildings, under the zinc roofs, where many people died. The zinc roofs, one of the very things that made the city so beautiful, so distinctive, were also one of the things that made the heat wave so lethal.

In 2015, Paris felt like a place of progress and victory. On the final day of the climate summit, leaders from 195 nations agreed to limit the Earth's warming to 2°C above preindustrial levels, which was — and still is — seen as the threshold of dangerous climate impacts. When French foreign minister Laurent Fabius banged his green gavel down in the cavernous conference hall to signal that the agreement was a done deal, the crowd cheered. Standing in the hall, I cheered too. For a brief, happy moment, it felt like the human race had finally come together and treated the climate crisis with the kind of urgency it demanded.

Like every other city in the world, Paris was built by people who believed that the Earth's climate was stable. Yes, there were hot and cold days, ebbs and flows of rivers, storms and droughts and other wild mood swings of Mother Nature, or angry gods, or physics, but the basic idea that there was a certain steady state and that the world would always return to it was never questioned. Just as no one built a city on the coast with the assumption that the polar ice sheets could melt and raise the water five or six feet in a few decades, no one built a city with the assumption that the temperature would jump five or ten degrees or that extreme heat waves would zap us. We built and lived in the Goldilocks Zone, and our cities are a part of that. They are Goldilocks cities.

But now, like everything else, these cities have to change. Cities on the coast have to adapt to rising seas. Cities in mountains have to adapt to raging rivers. And cities everywhere have to adapt to rising heat. It is the great urban engineering project of our time — making a city that was not designed for extreme heat into a city that is livable during extreme heat. Or, if that is too much, at least a city that is not a deathtrap for its citizens. And cities have to do all this while growing fast to handle the explosive urban population growth that is projected in the coming decades. By one account, the world will need to build an entire New York City every month for the next thirty years to accommodate all the new people.

Building a heat-friendly city is not an

impossible task. As David Hondula, now the chief heat officer in Phoenix, told me when we were driving around on a hot day, "If a city like Minneapolis can be designed in such a way that it's pleasant to live in during extreme cold, then the city of Phoenix can be designed in such a way that it can be pleasant to live in during extreme heat." At the time, I agreed with Hondula: there is certainly no reason why, if you have enough time and money, you can't engineer a city to be a safe and livable place at 130 degrees.

But the more I've thought about Hondula's remark, the more complex it has become. For one thing, it has *always* been cold in Minneapolis. The city was built in the cold and for the cold. There was no retrofitting necessary (and like all cities, it has its own challenges as the world around it heats up). That's a very different thing than taking a city that is built for one climate and modifying it to fit another. Just one example: basements in houses. In cold places like Minnesota, basements are invaluable for food storage, for infrastructure like furnaces, and for basic insulation of the home. It's pretty simple to dig a basement if you do it before you build a house. If you want to do it later, it's difficult and expensive.

The same thing with designing a city for

extreme heat. Maybe you'd want to put tunnels underground downtown so people can move around out of the heat. Maybe you'd want to arrange the streets so they funnel the breeze through the city. Maybe you'd build train tracks with steel that won't buckle when the temperature hits 120 degrees. Maybe you'd want to plant a heat-hardy tree in front of every house. All this stuff is much easier to do if you do it from the beginning.

For cities, the challenge of thriving on a superheated planet is twofold. First, as cities grow, how do you ensure that they grow in a heat-smart way? Another fifty years of suburban sprawl is not the answer. Cities need to be denser. Cars need to be replaced with bikes and public transit. New buildings need to be not only efficient and built of sustainable materials, but also safe for people during increasingly intense heat waves. That means more green space, more trees, more water, more shade, more thermally intelligent urban design.

The second, and more difficult, challenge is figuring out what to do with existing buildings and cityscapes. The vast majority of existing buildings are ill-suited for the extreme climate of the twenty-first century: poorly insulated, poorly sited, dependent on air-conditioning to keep them habitable.

Do you tear them down and rebuild? Do you retrofit? How do you create more green space in already crowded inner cities? How do you banish concrete and invite in nature?

In many cities, this urban remodeling project is already underway. In New York City, workers and volunteers have planted over a million trees to add shade and clean the air. In Seville, Spain, city planners are using the technology of ancient underground waterways to provide cooling for the city without depending on air-conditioning. In Freetown, Sierra Leone, officials are creating urban gardens, improving access to clean water, and erecting plexiglass awnings over outdoor markets to shade people when shopping. In Los Angeles, public works crews are painting streets white to increase reflectivity. In India, they are experimenting with green roofs, which absorb heat and create space to grow food. In Austin, my wife, Simone, transformed a useless plaza in front of the Blanton Museum of Art, where she is the director, into a public gathering space shaded by 40-foot-tall carbon fiber sculptures that rise like giant elegant flowers, creating a shaded microclimate with dappled light. Cities like Orlando, Florida, and Tempe, Arizona, are pioneering the development of resilience hubs — basically, community centers that are outfitted with back-up power, wifi, and air-conditioning, where residents can find refuge during extreme heat events (or other kinds of emergencies).

But perhaps nowhere in the world do the challenges, as well as the opportunities, loom larger than in Paris.

After the fifteen thousand deaths in France during the 2003 heat wave, the lesson most Parisians learned was that heat waves are deadly. It was not that climate change is making heat waves stronger, more frequent, and *more* deadly. "The basic response was 'We have to take better care of old people,'" said Franck Lirzin, a real estate executive who worked as an advisor to French president Emmanuel Macron. "Nobody made the connection with climate change."

According to Lirzin, who wrote an influential book about how Paris can adapt to climate change, Parisians have a long history of being ignorant about climate. In part, this is because Paris has always had such mild, equitable weather that no one has had to think much about it. "My wife is Dutch," Lirzin told me. "So I have spent a lot of time in Amsterdam. And it occurred to me that the Dutch have been thinking for hundreds of years about how to build houses that stay warm in cold weather and dry during floods. There is nothing like that in Paris." Similarly, architects and city planners in Marseille, on the Mediterranean coast, learned to deal with the heat by using thick walls, tile roofs that absorb the heat, orientation of streets to catch the cooling winds.

"For hundreds of years, we just built houses out of cheap stone and nobody thought about the climate, hot or cold," Lirzin says. "We have no climate culture in Paris, no history of thinking about it, no knowledge base to work from."

There are a lot of cities that have no climate culture. The Bay Area, where I grew up, has no climate culture because the weather was perfect almost all the time. Same with Mexico City, where my wife grew up. Other places have the inverse of a heat-savvy culture. People in Helena, Montana, where my mother and sister live, know very well how to deal with cold, but when it gets hot, they are clueless. They don't have air-conditioning, they don't have sunshades on their windows, they don't know that they should check on vulnerable friends and relatives. The reverse is also true: in Texas, people know how to handle heat. But when we get hit with a deep freeze, as happened during the winter of 2021, many people had no idea what to do. Should you drain your pipes? Should you leave your oven on to warm up the house? How the hell do you drive on ice without snow tires? (Answer: you don't, you stay home.)

Paris is celebrated today for its grace and beauty, but it wasn't always so. In the years after the French Revolution, Paris was a "worn, wrecked and exhausted city" that "smelt more of filthy mud and sewage than she had at the worst moments of the Middle Ages," one historian wrote. Behind the royal palaces was a city of dilapidated hovels, poverty, and prostitutes. In 1832, a major cholera epidemic, one of the worst epidemics in the city's history, killed 18,402 people.

Something had to be done. The man in charge was Louis-Napoléon, nephew of the first Bonaparte, who had terrorized all of Europe. Louis-Napoléon — who returned from years of exile to become president of the Second Republic briefly before being crowned emperor as Napoléon III — decided he was going to leave his mark by giving Paris a makeover. To mastermind his project, he chose Georges-Eugène (later Baron) Haussmann, a career bureaucrat and political ally. Haussmann had no training in architecture (he once described himself as "a demolition artist"), but he was a highly efficient, ruthless administrator and financial wizard.

Whatever the quality of his character may have been, Haussmann's vision for Paris ranks as one of the greatest urban achievements of the nineteenth century. In fact, it's hard to grasp what an enormous and brutal undertaking this was. At the time, Paris was home to more than a million people (twice the size of New York). Haussmann bulldozed through the medieval slums, displacing people by the thousands. He built parks and planted trees (over four hundred thousand trees and shrubs were planted in the Bois de Boulogne alone). He created wide boulevards and erected block after block of new apartment buildings with gray polished limestone façades and symmetrical wrought-iron balconies. They were the products of a new industrial age, mass-produced for the new bourgeois. The buildings were all five or six stories tall, with apartments getting less spacious and elaborate as you moved up. Privacy, hygiene, and comfort were priorities.

The "new" Paris was instantly controversial. One critic said the rows of buildings made him think of "some American Babylon of the future," while novelist Émile Zola raved about the transformation: "I love the horizons of this big city with all my heart . . . depending on whether a ray of sunshine brightens Paris, or a dull sky lets it dream, it resembles a joyful and melancholy poem. This is art, all around us. A living art, an art still unknown."

But the most notable thing about the buildings, from a climate perspective anyway, was Haussmann's use of zinc roofs. It was an innovation at the time, lighter and cheaper than tiles, corrosion-resistant, and virtually inflammable. As long as they are properly installed, the roofs last a very long time. As evidence of that, nearly 80 percent of the buildings in Paris today — more than a hundred thousand buildings — have zinc roofs. There is even a movement in France to get zinc roofs designated with UNESCO World Heritage status.

But the problem is, Parisians aren't living in the nineteenth-century climate anymore. In the twenty-first century, zinc roofs are deadly. On hot days they heat up like frying pans — literally. One researcher measured a temperature of 194 degrees on a zinc roof in Paris on a summer day. And because top-floor garrets were designed for servants and not insulated, that heat is transferred directly into the rooms below. During the 2003 heat wave, the poor ventilation and insulation of these garrets, as well as the difficulty for elderly or sick people to climb down six flights of stairs to escape, made them deathtraps.

So what's to be done with the zinc roofs now? "There are no good options," Lirzin said. What about adding insulation beneath the zinc? "That is very difficult," he explained. "The roof is not designed for additional weight, so it means removing and rebuilding the entire framework, which is very expensive." And just ripping off the roof entirely and replacing it with something better adapted to the climate of the twenty-first century would be, for many Parisians, an unthinkable defacement of their beloved city. "It would take years to get a permit, and then most likely you will be denied."*

Painting the roofs white could help. Light colors increase the albedo, or reflectivity, of buildings, deflecting sunlight and causing

* In 2021, the French parliament passed a law restricting the rental of poorly insulated houses and apartments. Since most apartments under zinc roofs are not insulated, most of these garrets will be pushed out of the rental market by 2035. This law was intended to protect people not from heat waves but from high electricity bills; however, in the end, it serves the same goal.

less heat to be absorbed (there's a reason why houses in hot places — Morocco, Portugal, the Greek islands — are traditionally painted white). White roofs can be remarkably effective in sunny climates. Researchers at Australia's University of New South Wales determined that white roofs reduced indoor temperatures by up to seven degrees. But since the zinc roofs in Paris are already light-colored, the impact would be more modest. There is also an issue with roof access and maintenance, since the white roofs would need to be repainted about every ten years or so. More significantly, white roofs are widely opposed by historic preservationists, who worry they would fundamentally change the look and feel of the city.

Green roofs are another possibility. In 2020, three young Parisians founded a company called Roofscapes to build what amounts to wooden platforms on top of the zinc roofs, which could become rooftop terraces. "People can grow food and get protection from the heat at the same time," said Olivier Faber, twenty-eight, one of the cofounders of the company, which first emerged as a start-up from the MIT School of Architecture and Planning. Faber points out that the wooden platforms are structured in such a way that the additional weight is borne not by the roof itself, but by the old stone load-bearing walls of the Haussmann buildings, which, in most cases, are plenty strong enough to support them. The idea was inspired by the rooftop terraces of Venice, a centuries-old tradition that Venetians built for access to fresh air and space to grow tomatoes. Paris officials have no problem with green roofs on new buildings — in fact, the city recently passed a law requiring them (or solar panels) on all new commercial structures above a certain size. It's the old buildings that are the problem. "It's taking people a long time to grasp the magnitude of what we face in Paris," Faber said.

This is not a Paris-only problem. The fight between the past and the future defines the battle lines in many cities struggling to adapt to our fast-changing climate. In Miami, preservationists want the great Art Deco buildings in South Beach to remain as they were in the 1930s when they were built; developers would be happy to bulldoze them and build condo towers that would be more resilient to hurricanes and flooding. In Venice, fifteenth-century palazzos are sinking into the lagoon, but they are such architectural treasures that it's impossible to imagine doing anything but spend hundreds of millions of dollars to shore them up for a few decades until they are reclaimed by rising seas. What's at stake here is not just architecture. It's our history, our culture, and our identity. But given the acceleration and urgency of the climate crisis, the harsh truth is, not everything can be saved.

It's not just zinc roofs that are untouchable in Paris. Exterior shutters, for example, can be effective shields against heat entering buildings. But if they were not included in the original Haussmann design, the historical commission prohibits building owners from adding them. "There is consensus that we have to do something to change these buildings to make them safer and more habitable in the future, but there is no incentive to find a solution," said Lirzin. "There is too deep of an idea that Paris is Paris and it can't change."

The cooling of Paris began in 2014, with the election of Anne Hidalgo as the mayor. Hidalgo, sixty-three, is the daughter of Spanish refugees who fled fascism. Her grandfather, a left-wing activist from Andalucía, was sentenced to death under the Spanish dictator Francisco Franco (he was ultimately spared). Hidalgo, who became a naturalized French citizen at the age of fourteen, started out as a factory inspector before becoming a government advisor under Prime Minister Lionel Jospin in the 1990s.

When Hidalgo took over as mayor of Paris, the once-great walking city was overrun with cars, air pollution in the inner city was deadly, bike lanes were rare, what few trees there were looked sickly. Hidalgo went after cars and trucks first, fighting what she said was a democratic war to give Paris back to Parisians. She closed two miles of roadway along the Seine in the center of Paris and turned it into a riverbank park. Rue de Rivoli, the city's main commercial street, became a boulevard for bikes and a limited number of authorized cars. In two of Paris's main squares, Place de la République and Place de la Bastille, traffic was squeezed to one side, making way for large pedestrian areas bustling with people. She created more than 250 miles of bike lanes in the city and was often photographed pedaling to her office in the Hôtel de Ville.

"My job is to transform this extraordinary, magnificent city without damaging it," Hidalgo has said. "To make it a city agreeable to live in but one that is a model that inspires, without denying its history."

But Hidalgo's ambitions got checked in 2018, when thousands of Parisians — many of them working-class people from the outer suburbs who commuted into the city — donned yellow vests and took to the streets to protest rising fuel taxes. The protests snowballed into a wider movement against French president Emmanuel Macron's supposed bias in favor of the elite and well-off city dwellers. Protestors set fire to barricades near the Arc de Triomphe, while police moved in with tear gas and rubber bullets to break up the crowds. The riots shook France and nearly toppled Macron's presidency.

After the riots, Hidalgo pivoted from cars to trees. She didn't entirely give up the fight against cars — in the run-up to the 2024 Olympics, she is pushing hard to ban most cars and diesel trucks from the inner city. But she discovered — as many politicians discovered before and after her — that fighting for trees is a much simpler battle. After all, who doesn't love trees? And Paris certainly needed more of them. Despite the city's many parks, it has one of the lowest tree canopy covers of any city in the world — only 9 percent, compared with 18 percent in Boston and 29 percent in Oslo. In the summer of 2019, Hidalgo launched an urban forest campaign, vowing to "significantly green" schoolyards around the city as well as four emblematic sites: the Hôtel de Ville, the Gare de Lyon, the square behind the Opéra, and a lane on the banks of the Seine.

From a public relations point of view, Hidalgo's urban forest announcement couldn't have been more timely: on July 25, 2019, less than a month after the urban forest initiative was launched, Paris recorded its highest temperature ever: 108.7 degrees (42.6°C). What better, more inoffensive way to cool the city than to plant a tree?

Trees are superheroes of the climate fight. They inhale CO_2 and exhale oxygen, filtering out air pollution with each breath. They suck up water from the ground and sweat it out through their leaves, which cools the air (think of them as mini–air conditioners). And of course they provide shade to all creatures great and small, as well as to the soil around them, which helps to reduce water loss through evaporation. As anyone who has taken a walk through a city park knows, they also offer mental health benefits to stressed-out urbanites. Trees are our deep-time evolutionary companions, fellow living things that we have spent millions of years leaning against, climbing, and worshipping.

As part of Hidalgo's urban forest initiative, the city of Paris plans to plant 170,000 new trees by 2026. That may sound like a lot. And in some ways, it is. But let's put it into perspective. New York City has planted over a million trees and is still going. Milan's urban forest project is planting 300,000 trees a year, with a goal of three million new trees in the city by 2030. Just to give you a sense of what this means on a global scale, there are about three trillion trees on the planet — which works out to about 422 trees for every person on Earth. Humans are responsible for the loss of fifteen billion trees a year. About five billion new trees are planted or sprout annually, yielding a net loss each year of ten billion trees. So as much as people may love trees, in the big picture, we are not very good to them. Since the beginning of human civilization, the number of trees on the planet has dropped by 46 percent.

Still, 170,000 trees are 170,000 trees. And when it comes to cooling off a city, trees matter. During the summer of 2022, one researcher found that on a hot afternoon the temperature on the ground in front of the Paris Opera House measured 133 degrees. A few steps away, under the shade of the trees on Boulevard des Italiens, the temperature on the sidewalk was only 82 degrees.

But in a rapidly changing climate, trees are not a simple answer to urban heat. For one thing, it's much easier to plant a tree than to keep it alive. People love to donate money to plant trees, and politicians love to get their pictures taken planting a tree, but it's much harder finding money for maintenance. In Los Angeles, city officials estimate that it costs $4,351.12 to plant and maintain a single oak tree for five years.* Then there is the question of who's responsible for taking care of them. The city of Phoenix, for example, does not have any centralized department or agency that is tasked with the maintenance of trees, which means that they often don't get the care and attention they need, especially in the first few years after they are planted. According to one tree advocate in Phoenix, the average life expectancy for a street tree in the city is only seven years.

"There used to be a lot of nice big shade trees in Phoenix, but they cut them all down in the 1960s because they were worried about how much water they used," Mark Hartman, Phoenix's chief sustainability

* Tree maintenance costs can be reduced by planting climate-appropriate trees. In parts of Arizona, for instance climate-appropriate trees like mesquite or ash only require extra water for the first year or two after they are planted. When they get bigger, the larger shade canopy often increases soil moisture by reducing evaporation.

officer, told me when I first visited the city a few years ago. In 2010, as the problems of extreme heat became more obvious, Phoenix officials set a goal of doubling the percentage of the city covered with tree canopy, from 12 to 25 percent, by 2030. But then came the inevitable budget cuts and layoffs. According to Hartman, "tree planting was cut back to stay only slightly ahead of those lost to storms and drought." Today, the tree canopy cover in Phoenix remains virtually unchanged from what it was a decade ago.

Even when they are properly cared for, city trees have a tough life. Dogs pee on them. Their roots are covered by asphalt and concrete. Lovers carve their initials into their bark. Drunk drivers run into them. In Athens, an invasive beetle is decimating the mulberry trees that provide shade in public squares. Ash trees, which make up the majority of shade trees in US cities like Chicago and Milwaukee, have been wiped out by the emerald ash borer, an Asian jewel beetle that arrived in North America in the early 2000s. One study found that emerald ash borers could kill 1.4 million street trees in the US by 2050. Here in Austin, big oaks and pecans are regularly cut down by tech bros making room for big houses and swimming pools. Laws and regulations don't stop them — so what if they have to pay a little fine? One of the most beautiful and historic trees in Austin, a six-hundred-year-old oak known as the Treaty Oak because it is where city founder Stephen F. Austin supposedly met local Native Americans to negotiate and sign Texas's first boundary treaty, was poisoned in 1989 by a guy who read some witchcraft books and thought killing the tree would somehow end his sorrow over being rejected by a woman (the tree was badly damaged but it's still alive).

Deciding which trees to plant is not a simple matter either. To protect against wide spread losses from disease and invasive species, diversity is important. But the climate of cities of today will not be the climate of cities in 2050. Arborists and urban planners find themselves casting forward, looking at which trees might be best suited to future conditions. In central Paris, the ubiquitous London plane trees are goners, vulnerable in a warming climate, and are being replaced by evergreens, oak, and buckeye. In Tucson, palm trees are out and paloverde and mesquite are in. To help people select plants that will survive in a harsher climate, researchers at Macquarie University in Australia developed a program called Which Plant Where. Home gardeners can go online, type in their location, and get recommendations for plants that will do well in the climate of, say, 2040. Already the warming climate is taking its toll on trees in hotspots around the world. When I was in Melbourne in early 2020 to look at the city's urban forest initiative, I took an afternoon walk through the Royal Botanic Gardens and came across a giant white oak lying across the grass like a fallen god. There was a hastily erected fence around the tree with a sign attached. It read, in part:

> Climate change is affecting the plants we grow — over the next 50 years, 20-50% of current plant species in botanic gardens and urban landscapes will likely be subjected to temperatures never experienced before. Last month, Victoria closed on one of its hottest and driest years on record, and the majestic White Oak (Quercus alba) that stood proudly and gloriously shady on Oak Lawn at Melbourne Gardens for over 150 years simply collapsed, leaving a hole in the city's most iconic landscape.

And it's not just old trees that are vulnerable. In 2011, the combined stress of drought and extreme heat killed off 10 percent of the urban trees in Texas. Nearly six million city trees died in just a few months. In the coming years, it may get a lot worse. A recent analysis of the Global Urban Tree Inventory, a database that contains 4,734 urban trees in 164 cites, suggests that in a midrange climate warming scenario, three-quarters of urban trees are likely to die from a combination of heat and drought by 2050.

There is also the question of equity. The basic truth is: rich people get nice trees, poor people get weeds. Mexico City is a striking example. On a recent summer day in Polanco, a neighborhood near the center of the city, I wandered down Avenida Presidente Masaryk, past Hermès and Cartier stores beneath the deep and decadent shade of big jacaranda trees. It was a reminder of the city's rich history of public gardens, parks and tree-lined squares. But in the sprawling neighborhoods that surround the inner city, it's a different story. As my mother-in-law, who was with us on the trip and who lived much of her adult life in the city, out in the poor *colonias* — "where most people actually live," she said — it's a landscape of concrete and struggling little ash trees.

The River Oaks neighborhood in Houston, where many wealthy oil and gas executives live, is full of majestic trees; Gulfton, a neighborhood five miles away where thirty languages are spoken and is often described as the Ellis Island of Texas, is an asphalt desert. In LA, Beverly Hills is a wonderland of exotic trees; South Central, not so much. Melbourne has magnificent elms and gums in the city parks downtown, but if you take the tram out to the western edge of the metropolitan area, it's hard to find a green leaf anywhere. According to a study by American Forests, a nonprofit that advocates for healthy forests and ecosystems, the city of Austin has the most unequal tree cover of any urban area in the US. Our neighborhood, where humble 1940s homes are being torn down and replaced one by one by glassy McMansions with fourteen-foot ceilings and black roofs, is shaded with great oak and pecan trees. But if I ride my bike out to the east side neighborhoods, where many Black residents were pushed by racial zoning laws (aka redlining) in the early twentieth century, the trees are smaller, the sun is hotter, the temperatures higher. In Austin, as in many other cities, city officials and volunteers have launched tree-planting campaigns to try to even out the tree inequity, but it's going to take a long time to democratize shade.

Urban forests are one part of a broader goal that many New Urbanists have of bringing nature back into cities — rivers, creeks, parks, gardens, animals, entire ecosystems that have been paved over and pushed out by relentless concrete and asphalt sprawl. Seoul, South Korea, spent $900 million to remove a highway and restore the Cheonggyecheon Stream in the middle of the city, which not only opened up much-needed green space, but also cooled the neighborhood around the stream by as much as ten degrees. To bring reclaimed water into the city to irrigate parks and green spaces, Athens plans to renovate an aqueduct first constructed by Roman Emperor Hadrian in 140 AD. New York City built the High Line, an elevated walk on the west side of the city that offers a leafy escape from urban concrete. Curitiba, Brazil, sometimes celebrated as the greenest city on Earth, has more than fifty square meters of green space per person (in contrast, Buenos Aires has two square meters per person). "We designed room for nature," one Brazilian official said.

In some cities, designing for nature is itself deeply artificial. In Singapore, it's hard to find a single inch of the city that is in any way "natural." But since the 1960s, there has been a deliberate government-led effort to adapt the city to rising temperatures. The highways are canopied with lush trees, urban parks have been expanded, and thousands of sidewalk trees have been planted. As I wandered downtown on one recent visit, I felt like I was in a jungle, there were so many vines and plants hanging from windows. The Oasia Hotel, designed by WOHA, an architecture firm based in Singapore, is a twenty-seven-story tower of green, with aluminum mesh panels that allow climbing plants to grow on the building. The plants and aluminum act as a sunblock, absorbing heat and creating natural shade.

All the greenery certainly helps Singapore keep cool for the people who live there. But it's hard to argue that cities like Singapore, which have massive ecological footprints through their oil refineries and supply chains that stretch around the globe, actually contribute to the cooling of the planet. "Singapore can make itself into a garden because the farm and the mine are always somewhere else," writes Richard Weller, a professor of landscape architecture at the University of Pennsylvania. "I would call Singapore a case of Gucci biodiversity, a distraction from the fact that they bankroll palm oil plantations in Kalimantan, the last of the world's great rain forests."

The centerpiece of the remaking of Paris is the Champs-Élysées, the iconic boulevard that runs between the Arc de Triomphe and the Place de la Concorde. Named after the Elysian Fields — a mythical Greek paradise — the Champs-Élysées was initially designed by King Louis XIV's gardener, André Le Nôtre. It was first called the Avenue des Tuileries and was lined with elm and horse chestnut trees that cut through fields and market gardens. Later, in 1709, it was renamed the Champs-Élysées and extended, and it became a popular spot for picnics by the end of the century. It was an urban manifestation of seventeenth- and eighteenth-century thinkers like Descartes and Galileo, who were founders of the scientific method. French formal gardens took on geometric designs, featuring a central perspective, with open views over long distances, and were designed following new mathematical and optical tools. "In this respect, the Champs- Élysées can be regarded as one of the 'zero milestones' of Western modernity," one French urban historian writes. "Reflecting a vision of domesticated nature, the avenue became a showcase of progress."

The first time I visited Paris in the early 1990s, I was shocked by how tawdry and touristy the famous boulevard was. It was like Times Square but worse, because with the monuments and the rows of sad-looking trees, you could see that it had once been a grand promenade. Over the years, the decline of the Champs-Élysées worried some of the property and business owners along the boulevard enough to hire PCA-Stream, a leading Paris architecture firm, to reimagine the boulevard in keeping with Hidalgo's vision of a cooler, greener city.

The cofounder of PCA-Stream, Philippe Chiambaretta, sees cities in a different way than most people. To him, they are not a collection of objects and human beings, or a giant machine, but a giant, sprawling organism that is itself alive and always changing. "A city has a metabolic flow," he told me. "Things are being made, energy is going in and out, it is always growing — or dying." Chiambaretta, who is in his late fifties, wore a colorful scarf the day we talked and was as happy to riff on Parisian history as he was modern architecture. "Like other organisms, cities can be healthy and in balance, or unhealthy and out of balance. A hot city, like a hot child, is a sick city."

Chiambaretta and his team spent four years looking at the Champs-Élysées from every angle, using a cross-disciplinary approach that included anthropology, philosophy, physics, and economics. Eventually, his team came up with a $300-million plan that transforms the Champs-Élysées from a wasteland to what Hidalgo calls "an extraordinary garden." In the plan, several lanes of traffic are eliminated to make room for bike lanes and broader pedestrian paths. The black asphalt is removed and replaced with a lighter-color paving that will reflect away sunlight. Rainwater is captured and recycled. And more than a thousand trees are planted in open soil that allows tree roots to mingle ("We know now that trees talk to each other, and we made an accommodation for that," Chiambaretta explained). All in all, besides making the Champs-Élysées safer, greener, and more fun to visit, Chiambaretta estimates that the makeover would lower the sidewalk temperature in the area by more than seven degrees.

And that's the thing with cities. They may be superorganisms with their own metabolic flows, but unless you have an emperor like Napoléon III in charge or a power broker like Robert Moses, the urban planner who ruthlessly reshaped New York City in the mid-twentieth century, retrofitting takes time. And that's assuming the city has the money and stable political leadership. Lirzin calculates that if Parisians retrofitted the historic buildings at a pace of about 1 percent per year, it would take seventy-five years to insulate and update them all. The Champs-Élysées project, Chiambaretta says, is unlikely to get started until 2025 at the earliest and will take a decade to complete. "If all goes well, we could be done in 2035," Chiambaretta told me. And this is just for one (big) block of one boulevard.

But perhaps the biggest hurdle is the gap between the grand visions of architects and urban planners and the reality of what might actually get built. This is always the case on big public works projects with a lot of different stakeholders, but it's particularly true with big public works projects that challenge people's expectations of how a city should look and feel. Chiambaretta is already facing this as he tries to build political support for his Champs-Élysées project in Paris. "The preservationists look at the light-colored paving we want to put in the Place de la Concorde to make it less of a furnace, and they say 'No no no, you cannot change the stones on the street, that is what makes Paris Paris!'" Chiambaretta told me. "And the tree people say, 'No no no, you can't remove any of the trees that are already there, even if they are diseased. If you try to cut down one, we will tie ourselves to it and call in the media!' And so what will get built, how far we will be able to go with this, I don't know yet. We can save the future or we can save the past, but we can't do both."

Lirzin fears the same thing with rules and laws that make it impossibly difficult to modify any of the Haussmann buildings in the center of the city. "Because the heat is not going to stop soon, people will have to do something," Lirzin says. "And what they will probably do is the same thing that people are doing all over the world — buy an air conditioner and stick it in the window. And for Paris, that would be a disaster. It would increase the demand on the gird, which would increase the risk of blackouts. And it would be ugly."

There are other ways to cool off Paris. Lirzin points out that many public buildings already use a district cooling system, which circulates water chilled by being underground through pipes in the building. The system could be expanded to include other parts of Paris, as well as private homes. But that would be a major undertaking.

Buildings could also be retrofitted so they don't need artificial cooling at all. Lacaton & Vassal, a French architecture firm that won the Pritzker Prize in 2021 (the Nobel Prize of architecture), specializes in reimagining old buildings in new ways. One of their best-known projects took 503 apartments in an ugly, inefficient concrete government housing building in Bordeaux and transformed them into bright, airy, well-ventilated dwellings. And they did it cheaply, and without displacing any residents (in fact, residents didn't even have to move out during the remodeling). The project was largely paid for by the French government, which had a worthy commitment to improving the lives of people in the old building. Why not scale it up and undertake a Haussmann-style retrofitting of every old building in Paris?

It's not unimaginable. After all, France is spending $40 billion to expand the Métro line beyond the inner city, adding 125 miles of track (mostly underground) and building 68 new Métro stations that will make it easier for people who live in the distant suburbs to get into the city without driving. The expansion, known as the Grand Paris Express and scheduled to be completed in 2030, will take 150,000 cars off the road.

"The challenge now is to carry out a Haussmanian transformation in a short time, but without guns, with a democratic process," said Alexandre Florentin, thirty-six, a member of the Paris city council who represents the Thirteenth Arrondissement, which is home to the city's largest Asian immigrant population. Florentin is part of a growing number of young Parisians who see extreme heat as a mortal threat to the City of Light. It's not just the zinc roofs, he said. It's the fact that schools aren't insulated or air-conditioned and hospitals are poorly built for heat and the vast majority of Parisians in his district are uneducated about how to handle heat. Florentin fears the city is headed for an apocalyptic future: summer blackouts, overrun emergency rooms, food shortages, epic traffic jams as people escape the city, firefighters dying of heatstroke as they fight wildfires in the Bois de Vincennes. "We have entered a new climate and energy paradigm," Florentin argued. "We need a social and cultural transformation on a level that I'm afraid people who have been in power for the last twenty years cannot really imagine."

How does that transformation happen? "You have to build a political movement," Florentin told me. "The people have to demand it." And it's not like the status quo is an option. One way or another, Paris — like every other city in the mid-latitudes — is going to be reshaped by extreme heat just as surely as it has been reshaped over the centuries by war and disease and commerce.

Florentin pushed the city council to establish a fifteen-member commission called "Paris at 50°C," which will hold public meetings around the city and make recommendations to the full council about the best strategies to adapt to extreme heat. "Here in Paris, there are three options," Florentin said bluntly. "We roast, we flee, or we act."

14. The White Bear

"Hey, guys, we have a visitor," Geoff Holmes called from outside the tent. "You might want to get up and check this out. And bring the gun."

We were high in the Canadian Arctic, on Baffin Island,* about halfway through a two-hundred-mile cross-country ski trek. I was traveling with two friends, David Keith, a professor of applied physics at the University of Calgary (he has since moved on to the University of Chicago), and Geoff Holmes, a Canadian engineer and former river guide who had rafted down some of the wildest

* Baffin Island is named after English explorer William Baffin, who sailed through the region in 1616 while searching for the Northwest Passage. The Inuit, who lived and hunted on the island for thousands of years before Baffin's arrival, refer to the region by its traditional Inuktitut name, *Qikiqtaaluk*.

rivers in Canada. In the first two weeks of our journey, which involved hauling plastic sleds loaded with a hundred pounds or so of camping gear, food, and whiskey across the ice, we had seen no one but a few wild-eyed BASE jumpers preparing to leap off a four-thousand-foot granite cliff in a fjord near the coast.

Nor had we seen any polar bears. That was a little surprising, given that Baffin Island has one of the densest population of polar bears in the world. Before we left Clyde River, the Inuit village where we began our trip, one of the Inuit elders had told us to be on our guard. Because of the unusually warm spring weather, he explained, "we've been seeing a lot of bears. They're on the move."

And so they were. From the tone of Holmes's voice, I knew exactly who — or what — our visitor was. I pulled on my boots and grabbed the twelve-gauge shotgun that was lying beside my sleeping bag. In any other situation, sleeping with a loaded shotgun was not my idea of fun. But in the Arctic, it was comforting to have it around. Sometimes, as we were about to drift off to sleep, Holmes would joke, "Is your girlfriend all tucked in and cozy?"

Keith and I stumbled out of the tent. It

was about 11 p.m., but because we were in the Arctic, it was not dark. At night, instead of sinking below the horizon, the sun just dipped low in the sky, creating a long, cold twilight that cast everything in a Hollywood glow. Holmes stood about twenty feet from the tent, where he had gone to take a piss. Several hundred feet beyond him was our visitor — a female polar bear, with a young cub close at her side.

It's a strange feeling, seeing a polar bear in the wild for the first time. Polar bears are such familiar creatures — they're on Coke cans and ice cream packages, and white teddy bears are on the beds of half the kids in America. And who hasn't seen *National Geographic*–style photos of a mama polar bear and her cubs trekking across the ice and not thought, *They're so cute!* In fact, media images of polar bears are so relentlessly charming that it's easy to forget that they are also wild animals, that, given the opportunity, will not hesitate to eat you. At least, I had forgotten that. But I remembered now.

We stood in silence for a moment, watching her. Her fur had a yellowish hue against the blue-white ice. She stopped and swayed her head slowly side to side, then lowered her nose so it was almost touching the ice.

Her black eyes and nose made an inverted triangle in her face. Her cub looked worried and huddled beside her.

"She looks hungry," Holmes said.

"Yeah," Keith said. "Let's hope she doesn't come any closer."

I was no polar bear expert, but I knew that a mother bear with her cub could be unpredictable and dangerous.

"Let's make some noise," I said.

"The gun is ready to go, right?" Keith asked.

I lifted the gun up to show him. It was loaded with three bear slugs, which are basically big hunks of lead that can stop a tank at twenty yards. I flicked the safety off. I was aware of myself thinking: *Are you really going to shoot a bear?*

"Yeah, ready," I said.

We started waving our arms and yelling as loud as we could. The bear understood immediately that we were trying to communicate with her. She reared up on her back legs. As she stood there, tall and erect and well balanced, she looked startlingly dignified and human. She lifted her nose, sniffing the air for clues.

"Shall I fire a warning shot?" I asked.

"Let's wait and see if she moves closer," Keith said.

She didn't. Instead, she dropped back

down on all fours and began to walk perpendicular to us. It wasn't a retreat, but it wasn't an advance either. The cub followed, straggling along in an adolescent way. We watched her for about a half hour, circling wide along the horizon.

"I think she's leaving," Holmes said hopefully.

We watched her lumber along for a while, growing smaller in the distance. I still couldn't quite grasp the fact that I was seeing a polar bear in the wild. It was like glimpsing Beyoncé cruising down Sunset Boulevard in a Bentley — the sight was at once surreal and predictable.

After the adrenaline drained away, I suddenly felt tired and sleepy. We had had a particularly difficult haul across the ice that day. And if I'd learned one thing on this Arctic trip so far, it was this: in a contest between fear and fatigue, fatigue usually wins.

We climbed back into the tent, joking about whether polar bears prefer North Face or Patagonia as an appetizer. Two minutes later, I was fast asleep.

The next morning, when we packed up for the day's journey, we noticed something alarming: fresh bear tracks thirty feet from our tent. While we slept, she had come back to investigate.

"Look at the size of those tracks," Holmes said, kneeling down for a closer look. Her paw prints were the size of pie plates.

It was a warm spring on Baffin Island. We were out on the ice for the entire month of May. There were plenty of days when the cold wind bit my face and the temperature dropped below zero at night. But on some days, we skied with our shirts off. Around us, ice-entombed mosses were thawing out for the first time in at least forty-five thousand years, suggesting, one researcher said, that temperatures are now warm enough "to melt all the ice in the eastern Canadian Arctic." As we skied along, we found that many glaciers were a mile or so farther back than where they were marked on our 1950s-era topographic maps.

Heat in a supposedly cold place is terrifying. Ice is a precision thermometer, registering the most minute changes. I felt it every day beneath my skis: on cold days the ice was hard and we could skate along, but if the temperature rose just a few degrees, it turned to slush and it was a slog to get anywhere. It changed every day, and every hour. On warm days, skiing over ice-covered fjords, I could hear cracks and feel the sea ice begin to buckle. Leads opened in the sea ice, and as long as they were less than a foot or so across, we could ski over them. But it was alarming to look down and see water beneath us.

For polar bears, heat equals starvation. They depend on sea ice to hunt seals. When the sea ice is gone, they can no longer hunt. Ideally, they gorge themselves on seals in the spring and early summer, then fast for the rest of the year, surviving on stored energy, or, if they are lucky, a whale carcass that happens to wash up on shore. In May, when we were on Baffin, the bears are ravenous for food. They know, somewhere in their inner bear-ness, that the season is moving on, that it is now or never. They must eat or die.

Pictures of hungry polar bears, trying to survive on rapidly melting ice, are among the most familiar and gut-wrenching images of our rapidly warming world. In 2018, a video of a starving polar bear on Baffin Island went viral, viewed more than two million times by people all over the world, eliciting outrage and sympathy. Paul Nicklen, the *National Geographic* photographer who shot the video, told the *New York Times* that the image of the staggering, skeleton-thin bear "rips the heart out of your chest."

Some scientists and climate activists balked

at the attention the polar bear video received, warning that emotional reactions are fleeting, and that it is the human impact of climate change that we should be focusing on, not suffering bears. But other scientists understood that telling stories about what's happening to polar bears is a useful bridge to helping people understand what's happening to human beings. "We care about the polar bears because they're showing us what's going to happen to us," says Steven Amstrup, the chief scientist for Polar Bear International, a nonprofit polar bear conservation group. "If we don't heed their warning, we're next."

The Arctic first appeared in the Western imagination in 330 BCE, when a Greek geographer and explorer named Pytheas left what is now the city of Marseilles and sailed for the Far North. It's not clear exactly what landmass he reached — it might have been Iceland, it might have been Greenland. Whatever it was, it lay six days north of England and one day south of what Pytheas described as a frozen ocean, a place that man could "neither sail nor walk." As one literary critic put it, "at a time when Aristotle was still hanging out in the agora, Pytheas had discovered pack ice."

Pytheas called the place he encountered Thule, as in ultima Thule — the land beyond all known lands. That is one of several names the Greeks gave us for the Far North. Another is Arctic, from *Arktikos* — "of the great bear." The reference was not to polar bears, which were unknown in Europe until the eleventh century, but to Ursa Major, the most prominent circumpolar constellation in the northern skies.

By the nineteenth century, the Europeans and the British were fairly obsessed with the Arctic. Much of the fascination was driven by the search for the Northwest Passage, which was a long-dreamt-of shorter water route between Europe and Asia whose discovery, it was hoped, would dramatically accelerate global trade. Others were obsessed with the glory of who could obtain "farthest north" — that is, the highest latitude yet reached by man.

And the obsession with the Arctic translated into an obsession with polar bears, which were shipped over in cages and became mainstays of traveling circuses and sideshows. The whiteness of polar bears was a subject of particular fascination. In *Moby Dick,* Herman Melville wonders why the color white, despite its association with things "sweet, and honorable, and sublime," strikes "panic to the soul." He uses the polar bear as an example, arguing that the bear's whiteness hides its "irresponsible ferociousness" with a "fleece of celestial innocence and love."

For hundreds of years, the North Pole was a destination for heroic Arctic explorers — American Robert Peary famously claimed to have reached the pole in 1909, although his accomplishment is now widely viewed as fraudulent. The first undisputed account of a human being setting foot on the North Pole was about British explorer Sir Wally Herbert, who, along with his team of dogs, arrived on April 6, 1969. By the time the centennial anniversary of Herbert's conquest arrives, because of the rapid warming of the region, you'll likely be able to travel across the Arctic in a sailboat.

But a vanishing Arctic is not just a symbol of our changing climate. It has real impacts on the lives of virtually everyone on the planet. A warmer Arctic alters the thermodynamic balance of the Earth's atmosphere, changing the pressure gradients that create heat waves, and altering rainfall patterns, especially in Europe and Asia, which will have big implications for food production. Rapidly melting ice sheets in the Arctic also accelerates sea-level rise, inundating coastal cities around the world, stranding billions of dollars' worth of real estate and forcing tens of millions of people to move to higher ground.

A warming Arctic is also speeding up the melting of permafrost, releasing vast quantities of methane, a greenhouse gas that is twenty-five times more potent than CO_2. More methane means more warming, which will release still more methane — when scientists talk about a looming climate catastrophe, this is one of the scenarios that worries them most. And it's not just methane and woolly mammoth bones that are locked up in the Arctic permafrost — there are also viruses and pathogens from an earlier time, which, as I mentioned in a previous chapter, when thawed and released into our world, could unleash a global pandemic (Bill Gates, an incurable optimist on many public health issues, once told me that pathogens released from thawed permafrost represent the one climate change impact that keeps him awake at night).

As the ice disappears, so will the polar bears. They have evolved to survive in a very particular ecological niche, one that is populated mostly by ice and seals. Their white coats blend in with the ice. Their front paws, exquisitely evolved by yanking seals out of the sea, are larger and shaped more like fish-hooks than the knife-like claws of other bears, which have evolved for climbing and digging as well as hunting.

Although polar bears can breed with grizzly bears, creating hybrid offspring, that is not a survival strategy for the species. Wildlife biologists believe the population of polar bears will dwindle with the ice, retreating farther and farther north. Polar bear rescue groups will likely arrange food drops from helicopters to keep the remaining bears alive (it's already a subject of active discussion). But as their population shrinks, so will their gene pool, leaving them more vulnerable to disease and less able to adapt to changing conditions. A few pampered polar bears will live on for a while in zoos, but unless we take radical action to stop heating up the climate in the next few decades, wild polar bears are goners.

That's what most scientists believe, anyway. But there's at least one scientist out there whose deep connection to nature and willingness to explore controversial ideas led him to wonder if there was a way to save the ice in the Arctic and, as a consequence, save the bears. And that scientist was David Keith.

■■■■

Keith is a tall, wiry guy with a narrow face and intense green eyes that seem to shoot laser beams when he is focusing hard on something. If you met him on the street, you might guess that he is a scientist. But you would probably not guess that he likes to spend his lunch breaks in a rock-climbing gym. It had long been his dream to ski across Baffin Island, and he did the lion's share of the organizing and logistical work to make our trip happen.

For Keith, the cold north was a landscape of adventure and beauty. Over beers on the deck of his home in Calgary, he told me about growing up in Ottawa, where one of his neighbors was Graham Rowley, an Arctic explorer who helped map Baffin Island in the 1920s. "His house was full of walrus tusks and he was always telling stories about encounters with polar bears," Keith recalled. "I wanted to grow up to be just like him." After college, Keith spent four months in a shack on Dundas Island in the high Arctic, studying walruses with Ian Stirling, a wildlife biologist who also happened to be one of the world's leading authorities on polar bears.

I had met Keith while I was reporting

a story about geoengineering — that is, large-scale manipulation of the Earth's climate to reduce the impacts of global warming — and heard that Keith was building a machine that could scrub CO_2 out of the atmosphere, which would then allow that CO_2 to be compressed and buried deep underground. A decade ago, when I met Keith, this was a radical idea. Today, the technology, known as direct air capture, is attracting billions of dollars of investment from companies like Google and Microsoft. Keith ended up starting a company called Carbon Engineering, which now employs 170 people and is building carbon capture projects for big corporations like Airbus.

At the time, Keith was also one of the leading thinkers in a far more outlandish proposal to cool the planet by using a fleet of high-altitude aircraft to spray sulfate particles high in the stratosphere. The particles would act as tiny reflectors, diverting a tiny amount of the sunlight that would otherwise hit the Earth. The idea more or less mimics the effects of a volcano eruption, which shoots sulfur into the atmosphere. The sulfur oxidizes into sulfuric acid, which accumulates in particles that drift around in the sky, reflecting sunlight. How much the particles cool the planet depends on how big the eruption is. When Mount Pinatubo, a volcano in the Philippines, erupted in 1991, it shot fifteen million tons of sulfur dioxide into the atmosphere and cooled the climate by about one degree for a year. A human-made geoengineering scheme would work in a similar way, and would create, to put it crudely, a thermostat knob for the Earth's climate.

"I think we have a moral obligation to take this idea seriously," Keith told me not long after we met. "I'm not saying we *should* do it, but even critics who think solar geoengineering is a bad idea agree that there is no technological or scientific reason why it wouldn't cool the planet. The big questions are, who benefits? Who suffers?"

There are a million reasons why this is a dangerous idea, including the fact that the particles rain out of the sky and so would have to be replenished every year or so, as well as the so-called moral hazard problem — if we can cool the planet by spraying particles into the stratosphere, why bother cutting fossil fuel pollution? Keith is hyper-aware of this, and is careful to underscore that solar geoengineering (also known as solar radiation management) is *not* a replacement for eliminating fossil fuels, but perhaps a way to take the edge off the heat until we can reduce emissions to zero.

One of the biggest concerns about solar geoengineering is how it might impact monsoons, which millions of people around the world depend on to bring water to the crops they need to survive. But the effect of dimming the sun on crops and shifting rainfall patterns is difficult to model (in fact, a number of papers have shown improved crop productivity). There is also some terrible death arbitrage to calculate in any discussion about geoengineering, since adding more particles to the atmosphere will undoubtedly lead to people inhaling them. Air pollution already kills as many as ten million people each year. However, Keith argues that air pollution deaths from the added sulfur in the air would be more than offset by declines in the number of deaths from extreme heat, which would be ten to a hundred times larger.

According to Keith, recent modeling of solar geoengineering suggests that the benefits would be particularly large for the poorest people in the hottest regions of the world. "That alone is a profound ethical reason to take this seriously," he told me.

But taking geoengineering seriously is a difficult thing for many people to do. The best way to stop heating up the planet is to stop burning fossil fuels and dumping more and more CO_2 into the atmosphere. As soon as that happens, the Earth's temperature will stop rising. Then, over the decades and centuries after that, as long as future humans don't go back to burning fossil fuels and dumping more CO_2 into the atmosphere, the Earth's temperature will slowly decline.

We would all like to think that's going to happen soon. But the unfortunate truth is that right now, the industrialized nations of the world are still dumping thirty-six billion tons or so of CO_2 into the atmosphere every year, which is roughly ten times faster than has ever happened in Earth's known history, even during past mass extinction events.

So that brings up the question: If some global consensus was reached that we really needed to cool the planet off fast, how would we do it? Are there tools we could use to take the edge off the heat? In a world of extreme heat, I wondered aloud one day on Baffin, what does a planetary air conditioner look like?

"A planetary air conditioner is not a great analogy," Keith replied. As we talked, we were resting on a slab of ice, with nothing but sky and rocks and more ice — always more ice — all around us. "If you have to make an analogy, a sunshade is better."

"Whatever you call it, it's still a dangerous idea. You're talking about spraying a bunch of particles into the sky."

"Yes," Keith said. "I'm suggesting we should not dismiss it out of hand. We should at least study it and learn more about the risks."

"I think most people would say that those risks are pretty big. You're talking about messing with the operating system of the entire planet."

"Okay. But we're already messing with the operating system of the entire planet in all kinds of ways. What do you think agriculture is? Every time you start up your car, you're dumping pollutants into the air and messing with the atmosphere. Humans are the dominant force on the planet, and it's our job to manage it as best we can."

Polar bears may look cuddly and cute, but they are the apex predators of the Far North. That is, there's nothing in the polar bear's realm that hunts *it.* It's afraid of nothing, because in its world, it has nothing to fear. Everything in a polar bear's domain is food until proven otherwise. Unlike other bears, which are omnivorous (grizzlies eat everything from plant roots to elk), polar bears are pure meat-eaters. While their preferred meal is seals, they have also been known to kill and eat walrus, beluga whales, and, in hard times, one another. Male polar bears have a particular appetite for cubs, which is why mother bears do everything they can to stay away from male bears when their cubs are young.

During the first two weeks of our trip, we spent most of our time winding through the fjords along the coast of Baffin, where the ice was solid. All the bears, we believed, were out on the floe edge — the place farther out where the open sea meets the frozen sea. For polar bears, this is the best hunting terrain, allowing them to catch seals when they pop up in their breathing holes in the sea ice, as well as when they haul themselves out to rest on chunks of ice adrift in the sea. Although we didn't know exactly where the floe edge was, we assumed it was miles from the shore. Our guess was that all the polar bears were busy hunting out there, and by hugging close to the shore of the island among the fjords, we were fairly safe.

But after the first sighting of the bear and her cubs, we learned otherwise. In the following days, we began seeing more and more bear tracks. Some were several days old; others looked like they had been made five minutes before we arrived. We crossed what we jokingly called bear highways, which were well-worn paths in the ice that bears used to travel back and forth from the floe edge. One day we discovered the solitary marks of a wandering bear that looped and circled for no apparent reason. But the farther we traveled, the more bear prints we saw.

Not surprisingly, we also sighted more and more bears. At least we thought we did. Usually it was just a glimpse in the distance, a moving shape that appeared against the dark background of a granite outcropping. But it was hard to be sure. Keith would be skiing along, and suddenly he would stop and gaze into the distance, and I knew exactly what he was doing.

"Anything?" I'd ask.

"Not sure," he'd say.

I'd get out my binoculars and look. Sometimes I'd see a bear lumbering along. Or what I thought was a bear. Maybe it was just a rock or a wisp of fog. The farther we went into polar bear country, the more often we stopped and looked. My brain was playing tricks on me, seeing bears where there were none, a manifestation of the creeping fear we were feeling as we moved deeper and deeper into the polar bears' territory. Were we being tracked? We didn't know.

The worst moments were when visibility dropped because it was snowing or overcast. On some days, we couldn't see twenty feet in front of us. One fogbound day, we crossed paw prints of a bear and her cubs that could not have been more than an hour old. So we knew they were near us. Was it the same family we'd seen before? We could see nothing. It was like being hunted by a ghost.

One day, when we came upon a towering iceberg trapped in the sea ice, Keith and I decided to climb it and get a view. Holmes thought it was an idiotic idea and kept a safe distance away — icebergs, which are really just big floating chunks of ice, are notoriously unstable and can flip over suddenly as ice melting below the surface alters their center of gravity. But Keith and I believed this particular iceberg was safely locked up in the sea ice and decided to go ahead. I gave the gun to Holmes. "Just in case we get a visitor," I said.

The iceberg was a hundred feet or so tall, with several flat outcroppings along the way. Keith climbed up one side, while I scrambled up the other. Using my ice axe to carve handholds and footholds in the ice, I pulled myself up to the first outcropping, which was a wide, flat area about fifty feet up.

As soon as I hauled myself up, I nearly

tumbled backward from shock — on the ice, right in front of me, was fresh bear scat. Fresh — as in, hours, or even minutes, old.

"There's a bear up here!" I yelled to Keith in a kind of panic. We both practically leapt off the iceberg.

When we met up with Holmes, who was waiting with the gun several hundred yards away, he looked at us both with astonishment.

"You guys are nuts," he said. "Do you *want* to die on this trip?"

One night a month or so before the trip began, Keith and Holmes and I sat in the living room of Keith's house and discussed our polar bear defense strategy. Keith explained that bear spray is useless in windy conditions. Some people who trek through polar bear country bring dogs to alert them of the presence of bears. We decided, however, that dogs were too much hassle. Among other things, they need to be leashed up at night, and you have to haul around a sled full of dog food.

Keith had a better way. "We'll use a bear wire," he said. It would work like this: every night after we pitched our tent, we would set our ski poles in the ground in a circle about thirty feet from the tent, driving them into the ground like fence posts. Then we would string a thin copper wire between them and connect both ends into a little box with a battery and a buzzer in it. The copper wire formed a circuit, and if a bear was coming into our camp, it would break the wire and the alarm would go off, giving us a few seconds' warning before it charged into our tent.

"And then what?" I asked.

"We yell and scream and throw things," Keith replied. "If that doesn't work . . . we shoot it."

We flew into Iqaluit, the capital of Canada's Nunavut territory, then jumped on a bush plane for a four-hour flight to Clyde River. It was disorienting to be in a landscape with no trees and nothing green of any sort. The village was a tangle of icy dirt roads and simple houses. Kids zoomed over the ice on ATVs. Before we left the village, I stopped in at the Northern Store — the only commercial enterprise in the village, as far as I could see — to purchase an extra box of bear slugs. We had one box with us, but I didn't want to take any chances.

For me, the first few nights on the ice were terrifying. I was convinced every gust of wind was a bear. But after a few days, my worry faded. Day by day, the sublime beauty of the icy world we were traveling through revealed itself to us. The towering granite walls of the fjords reminded me of Yosemite Valley (Keith and Holmes, both experienced climbers, mapped out various imaginary routes up the walls as we skied along). We were accompanied by a raven, which would appear and disappear along our route, speaking to us in spooky but almost intelligible vocalizations that suggested it had many stories to tell about life in the cold north.

Every day was its own little cocktail of adventure, a mix of pulled muscles, blisters, terror, tedium, and howling winds. One day, as we were skiing happily along, the ice gave way beneath me, and I suddenly fell into the water below. Luckily, we were near the shore and the water was only about two feet deep. We all laughed, although personally I wasn't sure how funny it was. I quickly pulled myself out and changed into dry pants and continued on.

Still, as the days passed, the risks of traveling in a rapidly melting Arctic became more and more obvious. When we were out farther from the shore, we would pass over cracks in the sea ice where we knew the water below was plenty deep. The biggest fear was not that we would fall in and drown, but that we would fall in and get soaked. If we didn't get out of the water fast, get the tent set up to shelter us from the wind, and get changed into dry clothes, hypothermia could set in. All we had for heat was a little single-burner backpacking stove. It could turn a pile of ice into a pot of boiling water, but it was hardly a roaring campfire to warm your bones.

We were all too aware that a hungry polar bear is a dangerous polar bear. For bears (as for other animals, including, for that matter, humans), hunger changes the calculation of risk. If a bear is on the verge of starvation, it's likely to be more aggressive about seeking out food. "Male bears, especially subadults, are usually the most aggressive polar bears," one polar bear biologist told me. "The only bear that is more dangerous for people out on the ice in the Arctic is a nutritionally stressed mama bear."

The risk of attack from polar bears is different than the risk of attack from grizzly and black bears. A substantial proportion of fatal attacks by grizzly and brown bears are defensive — you're hiking on a trail and turn a corner and there's a brown bear. It is as surprised as you are, and, feeling threatened, it attacks. Whereas polar bear attacks are predatory. They stalk their prey — whether it's a seal or a walrus or a human.

But bears, like humans, and like dogs and cats and every other higher animal, have personalities. There are cranky bears and mellow bears, curious bears and reckless bears, bears that were abandoned too soon by their mothers and bears that had healthy childhoods. Doug Peacock, a writer and outdoorsman who has spent years living among grizzlies near Glacier National Park, has attributed his survival to being able to know one bear from another, and to being able to tell, at a glance, which ones were dangerous and which ones were not.

As Keith and Holmes and I skied along on the softening spring ice, we were hyperaware of what the warming weather meant for the bears — if they didn't fatten up now, it was going to be a long and hungry summer.

We were also hyperaware that there was not much we could do to lower the risk of attack. While we were skiing, we were always watchful, scanning the horizon, looking behind us. We kept the shotgun handy. But it was at night, while we were asleep in our tent, that we were most vulnerable. All we had was the bear wire. I often stared up at the orange nylon roof of the tent, listening for sounds outside, thinking how fucking stupid it was for me to be lying out here on the ice in polar bear country with just a thin copper wire for protection.

Near the end of the trip, we saw bear tracks everywhere — old tracks, new tracks, cub tracks. It felt like they were all around us, closing in. But there was nothing we could do. Each night, we pitched the tent and strung up the bear wire, ate some oatmeal doused with olive oil (all the food we had left), played some Johnny Cash for an hour or so on the little solar-powered speaker we had brought with us, then double-checked the twelve-gauge and went to sleep. And each morning, when we woke up, we felt lucky to be alive.

We went a month without calling home, without email, without news, without social media (we had a satellite phone with us in case we needed to call out in an emergency, but it remained switched off for incoming calls). After we left Clyde River, we did not see another human being. We did not even see a *sign* of another human being — no footprints, no candy wrappers, no lost gloves. Skiing across the ice, we often skied in silence. I was as alone with the world — alone with nature — as I had ever been. I'd look at the sky and the ice and scan the horizon and think about how fragile it all was.

But I was not alone. Nor was I disconnected. I was skiing through a landscape that was in motion, shaped by the accumulated warming of two hundred years of fossil fuel burning, two hundred years of steam engines, coal plants, cars, trucks, ships, and airplanes. Two hundred years of electrification and land-clearing and cattle-raising and data-processing, two hundred years of what we all too easily define as "progress." Our world is a time machine, where history shapes the future, and going to the Arctic is no escape. If anything, in a remote place like this, you feel the connection even more deeply.

And I could see this world we were skiing through was vanishing fast. In a month or so, after the temperature warmed a few more degrees, the ice would all crack up and melt away. The bears had evolved to be part of this cycle, exquisitely adapted, with their white coats and their big strong seal-grabbing front paws. This cold world was their Goldilocks Zone. But if I learned anything on my trip to Baffin, it was that their Goldilocks Zone was disappearing fast. Just like ours.

After a month on the ice, we arrived at our last campsite. It was nothing more than a GPS spot on the ice where we had arranged to rendezvous with Inuit guides. From here, they would use snowmobiles to haul us the last five miles through an un-skiable mountain pass to Pond Inlet. However, when we skied up to the spot, we discovered that the Inuit guides had not yet arrived, and, more disturbing, that the meeting place was a polar bear slaughtering ground. Seal blood was splashed everywhere on the ice, along with seal flippers, tails, entrails. And among it all, enormous footprints of bears.

Keith and I looked at each other like *You've got to be fucking kidding me.* Holmes, as usual, found something funny to say. "God works in mysterious ways, doesn't she?" We tried using the satellite phone to call the Inuit guides to change the pickup location, but there was no answer.

We thought about skiing a few miles away and camping there, then returning to the rendezvous point in the morning. But we were exhausted. We had spent all our energy and most of our supplies to get to this point. The idea of skiing even another half mile was daunting to me. After two hundred miles of skiing, my body was worn out.

After discussing it for a few minutes, we agreed that we would ski five hundred feet or so off to the west, where there was a smooth patch of ice, and pitch our tent there. And then we would just be vigilant. We trusted that the Inuit guides would arrive in the morning.

So that's what we did. Was I terrified? Yes, until my mind shut down from fatigue. And then I felt nothing.

When I woke in the morning, I was thrilled to be alive. I stuck my head outside the tent and saw a beautiful bright blue sky. It was like Mother Nature was throwing us a party at the end of our trip. We cooked the last of our oatmeal and rolled up our sleeping bags for the last time. We decided to leave the tent up for shelter from the wind while we waited for the Inuits to appear with their snowmobiles. We went outside and fiddled with our gear and talked about how eager we were to see our loved ones back home and sleep in a warm soft bed. I was as happy as I'd ever been.

At about 10:30 a.m., we went back to the tent for a final celebratory cup of tea. Keith fired up the stove and we sat around in a circle. We had been through a lot and we knew we would never do anything like this together again. I had never felt closer to anyone outside of my family.

A few moments later, as I was blowing on my tea, watching the surface ripple, the bear wire alarm went off.

"The wind again," Keith said absent-mindedly.

Holmes nodded. "Well, this is the last time!"

I was closest to one of the tent doors. "I'll check it out."

I put my cup down carefully so I wouldn't spill the tea and turned around and unzipped the tent. The door was narrow, so I had to crawl out awkwardly. Just as I stood up, I saw her, fifty feet from me, a female polar bear moving aggressively toward the tent. Two young cubs trailed behind her.

As soon as she saw me, she stopped and reared up on her hind legs. She was much taller than me. I noticed how rough and dirty her underbelly was. Her cubs huddled beside her. She snorted and made an odd huffing sound. Her black eyes looked directly into mine, as if she knew exactly what was going to happen next.

But she did not move toward me. After staring me down for a few long moments, she shook her head the way dogs sometimes do when they get out of the water, then slowly dropped onto all fours. By that time Holmes and Keith had climbed out the other tent door and were standing near me. Keith pointed the shotgun across the tent at her. I knew if she took one step toward us, he would have no choice but to shoot her. Maybe she knew it too. She took a few steps away from the tent. But then, as if she were having second thoughts, she turned back toward us and reared up again and snorted and huffed. Again, she stared at us intently. And again, she dropped down and turned around and walked off, her cubs trailing behind her.

Watching her go, I felt like a guilty man who had gotten a last-minute reprieve. Whatever suffering this warm spring had brought to her and her cubs, whatever hardship they had endured from sharing a planet with another animal so hell-bent on melting her world, she decided not to take it out on us.

Epilogue: Beyond Goldilocks

There are no signs or border crossing guards at the edge of the Goldilocks Zone. If we cross over, no alarms will go off. Depending on where you live, you may cross over sooner than others. But unless we take dramatic action now, we may all discover what it's like to live outside the zone. The human race — which built the pyramids and the iPhone, wrote epic love poems and invented rock 'n' roll, worshipped ancient gods and now deifies Hollywood stars — will exist in a world beyond the world it grew up in, beyond the place where our hearts were shaped and our genes were forged. We will be, in the deepest sense, on our own.

Heat will be the engine of this transformation. The heat that propels us out of our Goldilocks Zone will not be accidental heat, the equivalent of an asteroid slamming into Earth. It will be deliberate heat. Premeditated heat. We plead guilty to first-degree heat, Your Honor. We have known for more than a century about the climate consequences of burning fossil fuels. And it wasn't just the scientists who knew. In 1965, President Lyndon B. Johnson was warned, as have been many presidents after him. By 1977, Exxon (now ExxonMobil) not only knew that decades of burning fossil fuels would heat up the atmosphere, but developed in-house climate models that projected those changes with remarkable accuracy. Despite that knowledge, we have not only continued burning fossil fuels, we have continued burning them with reckless abandon. In a sense, you could say we have built a heat-fueled rocketship that is taking us, for better or worse, on a trip beyond the Goldilocks Zone.

We are not there yet. It's hot, but the things that our ancestors lived with for millions of years — the deep forests, the cool ocean, the snow-capped mountains — are still with us, still recognizable, still our companions. Making the necessary changes will be hard; it will require political leadership and a deeper understanding of our connection with one another and with the world we live in. But it is not beyond our reach. "Humanity retains an enormous amount of control over just how hot it will get and how much we will do to protect one another through [the coming] assaults and disruptions," argues David Wallace-Wells, author of *The Uninhabitable Earth: Life After Warming.* The good news, Wallace-Wells points out, is that the world is decarbonizing faster than anyone anticipated a decade ago. And thanks to decades of innovation, clean energy is now cheaper than fossil fuel energy in most parts of the world. That means we now have the means to lift hundreds of millions of people out of energy poverty without relying on coal, gas, or oil. Right now, our dependence on fossil fuels is all about inertia, political will, and big oil and gas companies wanting to milk their investments as long as they can.

If we do end up pushing beyond our Goldilocks Zone, we'll be okay for a while. We have tools and technology to help us adapt and survive. At least the lucky ones have them. But our world will be transformed. The tree you used to climb when you were a kid will die. The beach where you kissed your partner will be underwater. Mosquitoes and other insects will be year-round companions. New diseases will emerge. Cults of cool will celebrate the spiritual purity of ice. You'll grill slabs of lab-grown "meat" and drink Zinfandel from Alaska. Your digital watch will monitor your internal body temperature. Border walls will be fortified. Entrepreneurs will make millions selling you micro cooling devices. Fourth of July celebrations will become life-threatening events. Snow will feel exotic.

In some regions of the tropics, outdoor life will become virtually impossible. People will flee, just like many other living things, to higher, cooler climates. In many parts of the world, survival will depend not just on access to clean water, decent food, and medical care. It will increasingly depend on access to cool spaces, a job that doesn't require you to work outside during the heat of the day, and the means to escape from extreme heat events if necessary. In those places, the lucky ones will look out the window at people fixing a power line or building a house in the heat with pity and sorrow and maybe a little fear. And if they have any self-awareness at all, they will see the distance between themselves and the people who sweat to keep our world functioning. Because the first and most striking consequence of the human race's trip out of the Goldilocks Zone will be the widening of the thermal divide, the invisible but very real line that separates the cool from the suffering, the lucky from the damned.

When it comes to imagining the future at the edge of the Goldilocks Zone, it's this thermal gap that is hardest to see. If the Covid-19 pandemic demonstrated anything, it was how quickly and easily people were able to normalize the deaths of others, especially if they were old, sick, or otherwise living on the margins. There were a thousand deaths a day from Covid in the US alone. There were headlines and speeches and heroic doctors and nurses. And if you lost a friend or loved one, you felt the tragedy of it all. But after the initial shock and fear of Covid, the deaths became a part of everyday life. Just as the 43,000 deaths a year in the US in auto accidents no longer trigger public outcry. Or the nine million deaths globally from air pollution each year. Or starvation in Yemen and Haiti. Or casualties of distant wars. It just becomes part of the world we live in.

And so it may be, I fear, with the suffering and deaths from extreme heat. It will become part of what it means to live in the twenty-first century, something we accept and don't think too much about in our everyday lives.

But the hotter it gets, the more difficult that will become.

Then again, maybe not. Maybe twenty thousand people dying in a sudden heat wave in St. Louis or New Delhi will spark a revolution. I met people while researching this book who believe that the political and economic systems that go with them are unsalvageable. You might be able to retrofit the buildings of Paris, they argue, but you can't retrofit the politics of Paris — or anywhere else, for that matter. The solution is to burn it all down and start over. So the sooner we get on with that, they argue, the better.

I've met others who believe that our neurological machinery is simply maladapted to the problems of modern life, especially in rich democracies like the US, where partisanship and political dysfunction reign and banning books is discussed with far more urgency than banning fossil fuels or educating people about the dangers of extreme heat. Hurricanes are wiping out cities on the Gulf Coast with ever more muscle, crops are failing, delivery drivers are dropping dead on the job on hot summer days and yet Matthew McConaughey is still doing TV ads for gas-guzzling SUVs. As one social critic puts it: "We are confronted simultaneously with our vulnerability to catastrophe and our profound unseriousness in the face of it. It's as if the fires are starting to spread through Rome and all we can do is argue about the fiddling."

In the long run, extreme heat is an extinction force. All life has a temperature limit, even the microbes that thrive in the thermal vents at the bottom of the ocean. Even things that aren't alive, like your phone or the server farms that power the internet, have limits. Some creatures, like some humans, are more fragile than others, but eventually, scientist James Hansen wrote, "the planet will quickly get on the Venus Express."

And you don't have to go to Venus to find evidence of the killing effects of a superheated planet. You can just visit Guadalupe Mountains National Park in Texas.

Guadalupe Mountains is one of the least-visited national parks in the country, a place that most scenery-seekers accustomed to the sheer awesomeness of Yosemite or Glacier probably view as not much more than a big pile of rocks in the desert. But what looks like desert today is actually an ancient seafloor. And what looks like a pile of rocks is actually the remains of a 260-million-year-old reef that grew in a vast inland sea that had once existed between the Gulf of Mexico and the Arctic. Where lizards live today, sharks once swam. The highest point of the old reef, a peak aptly named El Capitan, is also one of the highest points in Texas (elevation 8,085 feet). It looms over the desert like the prow of a giant ship.

Although El Capitan looks like a mountain, it's actually an enormous pile of skeletons from the creatures that once lived on this reef, a towering mass grave from a time on Earth when it was so hot that all the ice was gone from the poles and sea levels were hundreds of feet higher than they are today. It is also stark evidence that as hot as the climate is today, it can get much, much worse.

Simone and I drove up from Austin one fall day to explore the old reef. It was a long drive, punctuated by roadrunners dashing across the highway and glimpses of coyotes trotting through the creosote bushes. We finally pulled into park headquarters near the base of El Capitan, where we picked up trail maps and looked at fossil displays in glass cabinets. Just out of sight over the horizon were the wellheads of the Permian Basin, the biggest oil-and-gas-industry playland in America. The land there has been drilled and fracked and the remains of the ancient animals that swam in this sea have been sucked into pipelines and burned to power modern life, including the gasoline in the car we were driving. It was a reminder, if any reminders are needed, of the depth of our entanglement with fossil fuels.

We headed over to the Permian Reef Trail. The ranger at the station there was a twenty-something woman who was very sweet and talked about how she often hiked this trail with her father, who was a petroleum engineer. She told us that this was "the premier geologic hike in America." It was eight miles to the top of the reef.

Up the trail we went. We saw a type of rock called boundstone, which was formed by algae, growing around sponges. We found traces of worms that burrowed through the seafloor millions of years ago searching for food. We found fragments of clamlike creatures called brachiopods. And we found reef boulders nine feet tall.

As we hiked, I tried to imagine what the world looked like when this reef was alive. The Earth's landmasses were joined in one supercontinent called Pangea. For most of the Permian period, which began about 300 million years ago, it was a little cooler than it is today. There were trees and plants on land, including conifers that were not unlike pine trees today. Big sail-backed reptiles and meat-eating gorgonopsians — imagine a cross between a Tyrannosaurus rex and a saber-toothed cat — were top predators. Small creatures called cynodonts, which looked like scaly rats and are one of the earliest ancestors of today's mammals, scurried around. In the ocean, there were giant sharks and bony fish, as well as millions of trilobites on the seafloor and brachiopods of many sizes and shapes.

The Permian lasted about 50 million years. Then, suddenly, over a period of maybe sixty thousand years — the blink of an eye in geologic time — everything died. Or nearly everything died. What killed life in the Permian was a bolt of extreme heat, brought on by violent eruptions of volcanoes in Siberia, which dumped billions of tons of CO_2 into the atmosphere very quickly. The temperature of the Earth may have jumped as much as 60 degrees. One-hundred-forty degree heat waves on the land wouldn't have been unusual. In the tropics, the ocean hit 104 degrees, which is about the temperature of water in a hot tub. Enough lava erupted from these traps to cover the entire United States in molten rock a half mile deep. It would take life on Earth ten million years to recover.

The End-Permian extinction was a horrific event, beyond our capacity to imagine. It was mass death by heatstroke.

We think we're a long way from recreating Guadalupe Mountains here in the twenty-first century. But we are not. Despite all we have learned about the risks of life on a superheated planet, despite all our technological sophistication and our knowledge of history, we are still hiking up that same mountain trail, one that leads not just to a vista point atop of a pile of skeletons, but into the desert beyond the Goldilocks Zone. "Right now, in the amazing moment that counts to us as the present, we are deciding, without quite meaning to, which evolutionary pathways will remain open to us and which will be forever closed," Elizabeth Kolbert writes in *The Sixth Extinction: An Unnatural History*. "No other creature has ever managed this, and it will, unfortunately, be our most enduring legacy."

Writing this book has been an adventure to unexpected places. I now feel — or think I feel — vibrating molecules when I pick up my coffee in the morning. I find myself calculating the thermal intelligence of every building I enter. I see lakes and rivers as cooling centers. Asphalt parking lots feel like ruins from a lost civilization. I judge politicians on how well they understand how fast our world is changing. When people ask me if it's hard to write about the climate crisis and imagine the hardship and suffering to come, my answer is always the same: This is the great story of our time, one that I feel privileged to tell. And yes, it gets dark sometimes. But it is also endlessly inspiring because I meet so many people who are fighting for the future and reimagining everything about how we live on this planet. I introduced you to some of them in the previous pages. With their help, and the help of other people like them, I believe we can build a better world if we want to. But I know that is a simple thing to say and a hard thing to do and that "better" means different things to different people. There are no maps for this journey we are taking, no virtual reality tours of the road ahead. "How do we face the truth of what is at stake and how much there is to do?" asks marine scientist and climate activist Ayana Elizabeth Johnson. "How do we muster up the courage to not give up in spite of the odds? How do we focus on solutions and on what each of us can do to help turn things around?"

I can't answer those questions. But I do know that after working on this book for three years, I think differently about other living things who share the burden of heat with us. When I see a bat dart across the evening sky, I consider how lucky it is to be an animal that hunts at night when it's cooler. When I see an armadillo waddling across our driveway on a summer night, I wonder if it's wishing it could shed its heavy scales. When I see a pecan tree with browning leaves, I wonder if it is stressed by the heat and what it's saying about these hard times to other trees nearby. And I think about the polar bear that didn't eat us on Baffin Island and how her cubs (who, if they survived, are probably now parents or grandparents themselves) are dealing with the disappearing ice. Have they learned new hunting skills? Are they becoming climate-smart bears?

For me, the big surprise in writing this book has been discovering not only how easily and quickly heat can kill you, but what a powerful reminder it is of how deeply connected we are to one another and to all living things. Wherever we may be headed, we are all on this journey together.

ACKNOWLEDGMENTS

The book was born on a 117-degree day in Arizona and came of age during a long midnight drive across Texas with my wife, Simone, during which she convinced me that heat was an important subject for a book and I needed to write it. Now I have. But I could never have done it without the help and support of many people.

My agent, Heather Schroder, has guided me through many mud bogs and over many mountains. Reagan Arthur saw the importance of this book from the beginning and trusted that I could pull it off. At Little, Brown, I'm grateful to Phil Marino, Bruce Nichols, and especially my editor, Pronoy Sarkar, for pushing me — in the most inspiring sort of way — to write a book that would ask people to think differently about their world. Elizabeth Garriga is brilliant at getting people to pay attention. Thanks to Barbara Perris once again for making sense of my tangled sentences. And to Linda Arends for shepherding the manuscript through production.

Several chapters of this book are built around articles I wrote for *Rolling Stone,* my journalistic home for many years, where I am lucky to work with people who understand the urgency of covering climate change. I am indebted to Jann Wenner, Will Dana, Jason Fine, John Hendrickson, Phoebe Neidl, Noah Shachtman, Hannah Murphy, Cadence Bambenek, and especially my long-time editor and foxhole buddy, Sean Woods.

I've traveled a long way with colleagues at the Adrienne Arsht-Rockefeller Foundation Resilience Center, especially Kathy Baughman McLeod, a climate warrior like no other. I've been inspired by Mauricio Rodas and Eleni Myrivili, who have been with me on this journey since the beginning.

Thanks to Kishna Mohan, Ashwini Chidambaram, and Vanessa Peter, for making connections for me in a world where I had no connections. To my shipmates in Antarctica, especially Rob Larter, Alastair Graham, Kelly Hogan, Tasha Snow, Lars Boehme, James Kirkham, Bastien Queste, Guilherme Bortolotto, and Anna Wåhlin. To Captain Brandon Bell, Chief Mate Rick

Wiemken, and Third Mate Luke Zeller for getting us through the ice and back again. To the scientists who have been so helpful to me over the years: Michael Mann, Ken Caldeira, Zeke Hausfather, Andrea Dutton, Jason Box, Andrew Dessler. To my researchers, Lucy Marita Jakub and Elizabeth Morison, for helping me begin. To Toby Kent, for the walk around Melbourne. To Betsy Abell, for sharing her family history with me. To Marc Coudert for frank answers to complicated questions. To Andrew Grundstein, Daniel Vecellio, Jeffrey Ross-Ibarra, John Whiteman, Sam Cheuvront, Jill Pruetz, and Peter Kalmus for the expert reads.

I'm grateful to Dan Dudek, whose wisdom I have tapped with every book I've written. To David Keith and Geoffrey Holmes for keeping me alive on Baffin Island. To Eric Nonacs for the long friendship and editorial wisdom. To Mike Dugan for demonstrating what it means to be a great neighbor. To Dr. Karl Koenig for keeping me moving. And to my friend Russell Banks, who left us just as I was finishing this book, for proving that being a great man and a great writer are not incompatible roles.

I'm not sure how to express my love and gratitude to Mary and Gary Wicha for taking me in — and for being so cheerful about it! Nicole, Rene, Erik, Ulan, and Amil have warmed my world. My brave and generous mom, Arlene Wadlow, gave me everything. My sister, Jill, has been heroic at a time when a hero was most needed. Grace, Georgia and Milo are the reason I write and the future I hope for.

And finally to Simone — my muse, my first and best reader, my fellow traveler, my monster-wrangler. You are, and always will be, hot.

GLOSSARY

albedo: the ability of an object or surface to reflect away sunlight, and, along with it, reflect away heat. A white roof has a high albedo; a black asphalt street has a low albedo. Shifting to higher-albedo materials is a key strategy in counteracting the urban heat island effect. Cold-blooded animals like chameleons manipulate their albedo to control their body temperature: when it's hot, they turn a white-ish color to reflect away sunlight. When it's cold, they darken to warm themselves.

core body temperature: the temperature of a person's internal organs, including the heart, liver, brains, and blood. Core body temperature is different — and more indicative of hyperthermia — than peripheral body temperature, which is the temperature at or near your skin, and is influenced by your environment. A person's core body temperature has a daily circadian rhythm, causing it to vary by a degree or more depending on the time of day. Body temperature is typically lowest in the morning, rises during the day, then falls again in the evening.

Goldilocks Zone: the area around a star where it is not too hot and not too cold for liquid water to exist on the surface of planets. Also known as the habitable zone.

heat dome: an area of high pressure that lingers for days or even weeks, trapping hot air underneath — a bit like a lid on a pot. In northern latitudes, pressure systems usually move from west to east, but sometimes they get blocked, often when the jet stream weakens and buckles. The jet stream is a band of strong winds high above the Earth's surface that, among other things, helps to develop and move areas of low pressure around.

heat exhaustion: a condition of elevated heat stress. Symptoms may include heavy sweating, dizziness, nausea, and fainting.

heat index: a value calculated by combining relative humidity and air temperature. It was developed from a model created by physicist Robert Steadman in 1979 as a way to more accurately measure how weather conditions feel to a person (that's why the heat index is sometimes referred to as the "feels like" temperature). The National Weather Service (NWS) cautions that heat index values assume shady, light wind conditions — exposure to direct sunlight can increase the index by 15 degrees. David Romps, a climate scientist at the University of California, Berkeley, has demonstrated that at high temperatures the NWS heat index calculation becomes increasingly inaccurate and can underestimate the actual heat index by as much as 20 degrees.

heat stress: the inability of the body to rid itself of excess heat. Initial symptoms include moderate sweating and a rapid pulse. Escalating stress leads to heat exhaustion and heatstroke.

heatstroke: a life-threatening illness typically associated with an uncontrolled rise in core body temperature above 104 degrees (40°C) and central nervous system dysfunction, including delirium, convulsions, coma, and death.

hyperthermia: abnormally high body temperature, caused when your body is absorbing or generating more heat than it can release. It's the opposite of hypothermia, which occurs when your body is losing more heat than it can generate.

thermal barrier: the reduction of heat

transfer from one body or space to another. On mammals, fur is a thermal barrier. In buildings, insulation is a thermal barrier. The phrase is also used in the context of a heat boundary, such as the speed at which a rocket can fly before aerodynamic heating (caused by the friction of an object flying through the air at high speed) causes the metal surface of the rocket to deform or melt.

thermal comfort: a person's state of mind that defines whether they feel too hot or too cold. Unlike hyperthermia, which describes a physiological condition, thermal comfort describes a psychological state. To put it another way, what any given temperature feels like to a person is dependent on who is doing the feeling. Age, basic health, clothing, drug use, and many other factors can have a significant influence.

transpiration: the process of water movement through a plant and its evaporation from its leaves. Plants pull water from the soil through their roots and transport it to their leaves, where photosynthesis takes place. But almost all the water that reaches the leaves is lost as water vapor when the stomata (tiny pore-like structures on the leaf surface) open and close to exchange CO_2 and oxygen. As the water vapor evaporates, it has a cooling effect on the plant, similar to the way sweating works to cool the human body. Large trees can transpire as much as 30 gallons of water a day, which draws heat energy from around the tree and cools the surrounding air.

urban heat island: a populated area that is significantly warmer than its surrounding rural areas due to human artifacts and activities. Pavement, concrete, and steel all absorb heat and radiate it back. Buildings block cooling breezes. Fewer trees mean less shade and less transpiration cooling. Waste heat from machines — cars, trucks, AC units, power plants, factories — further adds to the heat load. The result can be temperatures 15 to 20 degrees hotter in cities than in nearby rural areas, with the temperature differential often largest at night.

vector-borne diseases: human illnesses caused by parasites, viruses, and bacteria that are transmitted by the bite of an infected arthropod, such as a mosquito or tick. Arthropods are all cold-blooded creatures and thus especially sensitive to changes in temperature and climate.

wet bulb globe temperature (WBGT): a measure of heat stress in direct sunlight, which takes into account temperature, humidity, wind speed, and solar radiation.

"Wet bulb globe temperature came out of the military, because they wanted to find a better way to prevent heat casualties," Andrew Grundstein, a professor of geography at the University of Georgia, told me. It's often used in sports, in the military, and for worker safety. "Because it considers so many factors, including the cooling power of sweat, it's a much more accurate measurement of heat stress than heat index," explained Grundstein. "The problem is, weathercasters shy away from it because it can give you a reading that is below the air temperature, so it confuses people." For example, as I write this, the temperature in Austin is 95 degrees, with a humidity level of 40 percent and a light wind. The heat index is 99 degrees. But the WBGT is only 86.4 degrees, partly because there is light cloud cover, and partly because humidity is relatively low (for Texas).

wet bulb temperature (WBT): a calculation of the lowest temperature that the evaporation of water can cool air. It was developed in the early twentieth century by British physician J. S. Haldane, who descended into hot, humid Cornish tin mines to research the impacts of moist heat on workers. Haldane developed a temperature measurement that focuses on humidity and how well someone can cool via the evaporation of sweat.

The phrase "wet bulb" refers to the fact that it is measured by wrapping a thermometer in a wet cotton rag or sock and placing it outdoors. As the water evaporates from the rag, it lowers the temperature of the thermometer, approximating the cooling capacity of sweat. When the air is dry, more water can evaporate, lowering the WBT temperature; when it's humid, evaporation is less effective, so the WBT is higher.

Researchers sometimes use the WBT to establish thermodynamic limits on heat transfer by the human body.

WBT is different from WBGT (see previous page), which was developed much later and includes the effects of wind and solar radiation.

NOTES

Heat Index

30 million: Luke Kemp et al. "Climate Endgame: Exploring Catastrophic Climate Change Scenarios." *Proceedings of the National Academy of Sciences* 119, no. 34 (2022), e2108146119. https://www.pnas.org/doi/10.1073/pnas.2108146119

2 billion: Ibid.

1 mile per year: Colin Carlson et al. "Climate Change Increases Cross-Species Viral Transmission Risk." *Nature* 607 (2022), 555–562. https://doi.org/10.1038/s41586-022-04788-w

2.5 miles per year: Ibid.

210 million: "Global Food Crisis." World Food Programme website. Accessed October 2022. https://www.wfp.org/emergencies/global-food-crisis

21%: Ariel Ortiz-Bobea et al. "Anthropogenic Climate Change Has Slowed Global Agricultural Productivity Growth." *Nature*

Climate Change 11 (2021), 306–312. https://doi.org/10.1038/s41558-021-01000-1

250,000: Meghan Werbick et al. "Firearm Violence: A Neglected 'Global Health' Issue." *Global Health* 17, no. 120 (2021). https://doi.org/10.1186/s12992-021-00771-8

489,000: Qi Zhao et al. "Global, Regional, and National Burden of Mortality Associated with Non-Optimal Ambient Temperatures from 2000 to 2019: a Three-Stage Modelling Study." *The Lancet Planetary Health,* vol.5, issue 7 (July 2021), 415–425. https://doi.org/10.1016/S2542-5196(21)00081-4

Prologue — The Goldilocks Zone

Jennifer Vines: James Ross Gardner. "Seventy-Two Hours Under the Heat Dome." *The New Yorker,* October 11, 2021. https://www.newyorker.com/magazine/2021/10/18/seventy-two-hours-under-the-heat-dome

76 degrees to 114 degrees: Personal communication with Portland office of National Weather Service, October 2022.

trees screaming: Bob Berwyn. " 'We Need to Hear These Poor Trees Scream': Unchecked Global Warming Means Big Trouble for Forests." *Inside Climate News,* April 25, 2020. https://insideclimatenews.org/news/25042020/forest-trees-climate-change-deforestation/

Many jumped: Hannah Knowles. "'Hawkpocalypse': Baby Birds of Prey Have Leaped from Their Nests to Escape West's Extreme Heat." *Washington Post,* July 17, 2021. https://www.washingtonpost.com/nation/2021/07/17/heat-wave-baby-hawks/

body bags with ice: JoNel Aleccia. "As Extreme Heat Becomes More Common, ERs Turn to Body Bags to Save Lives." *Kaiser Health News,* July 22, 2021. https://khn.org/news/article/killer-heat-body-bags-ice-heatstroke-emergency-treatment-climate-change/

Shandas recalled: Personal communication with the author, October 2021.

99 degrees: Ibid.

official count was 1,000: Kristie L. Ebi. "Managing Climate Change Risks Is Imperative for Human Health." *Nature Reviews Nephrology* 18 (2021), 74–75. https://doi.org/10.1038/s41581-021-00523-2

Rosemary Anderson: Jaelen Ogadhoh. "14 in Clackamas County Die So Far in Summer Heat Waves." *Canby Herald,* August 10, 2021. https://pamplinmedia.com/wlt/95-news/518067-413985-14-in-clackamas-county-die-so-far-in-summer-heat-waves

Jollene Brown: Gardner, "Seventy-Two Hours Under the Heat Dome."

How strange: Vjosa Isai. "Heat Wave Spread Fire That 'Erased' Canadian Town." *New York Times,* July 10, 2021. https://www.nytimes.com/2021/07/10/world/canada/canadian-wildfire-british-columbia.html

cat in a cage: Norimitsu Onishi. "After Deadly Fires and Disastrous Floods, a Canadian City Moves to Sue Big Oil." *New York Times,* August 29, 2022. https://www.nytimes.com/2022/08/29/world/canada/vancouver-floods-fires-lawsuit.html

"Ten minutes later": Cathy Kearney. "B.C. Man Says He Watched in Horror as Lytton Wildfire Claimed the Lives of His Parents." *CBC News,* July 2, 2021. https://www.cbc.ca/news/canada/british-columbia/son-recounts-horror-of-losing-parents-in-lytton-bc-fire-1.6088297

a billion sea creatures: Valerie Yurk. "Pacific Northwest Heat Wave Killed More Than a Billion Sea Creatures." *E&E News,* July 15, 2001. https://www.scientificamerican.com/article/pacific-northwest-heat-wave-killed-more-than-1-billion-sea-creatures/

extreme cold kills people: Comparisons between deaths from extreme cold and deaths from extreme heat are not easy to make. First, according to Kristie Ebi, an epidemiologist at the University of

Washington, evidence that people die from heat is well-documented. "But research that people actually die from the cold is scattered and not strong," she argues. Many studies have shown that cardiovascular disease increases in the winter, but disentangling cold temperatures from seasonal factors is challenging. "During the winter, on average, blood pressure, blood viscosity, and cholesterol go up," explains Ebi. "But we don't know how much could be due to temperature separate from behavioral changes, day-length, and other factors." Second, comparisons of cold mortality vs. heat mortality often compare deaths during winter to deaths during heat waves, which is apples to oranges (winter is a season, a heat wave is a temperature event). Third, as the world heats up, all projections indicate that heat-related mortality will rise — key questions include who and where it will kill. As for the notion that, in a warming world, fewer cold deaths somehow offset increased heat deaths: "At its core, this argument implies that it's acceptable for Aunt Harriet to die from the heat because Uncle Joe wouldn't die from the cold," says Ebi. "This argument doesn't consider the moral implications at the individual level."

"Hot or Not": Elaina Dockterman. "How 'Hot or Not' Created the Internet We Know Today." *Time,* June 18, 2014. https://time.com/2894727/hot-or-not-internet/

"the elemental heat of the semen": Danielle Jacquart and Claude Thomasset. *Sexuality and Medicine in the Middle Ages* (Princeton, NJ: Princeton University Press, 1988), 59.

900 million people: Dennis Wong and Han Huang. "China's Record Heat Wave, Worst Drought in Decades." *South China Morning Post,* August 31, 2022. https://multimedia.scmp.com/infographics/news/china/article/3190803/china-drought/index.html

"world climatic history": Quoted in Michael Le Page, "Heatwave in China Is the Most Severe Ever Recorded in the World." *New Scientist,* August 23, 2022. https://www.newscientist.com/article/2334921-heatwave-in-china-is-the-most-severe-ever-recorded-in-the-world/

Heather McTeer Toney: "Making the Case for Climate Action: The Growing Risks and Costs of Inaction." Testimony before the House Select Committee on the Climate Crisis, April 15, 2021. https://docs.house.gov/meetings/CN/CN00/20210415/111445/HHRG-117-CN00-Wstate-McTeerToneyH-20210415.pdf

Researchers at Dartmouth College: Christopher W. Callahan and Justin S. Mankin. "Globally Unequal Effect of Extreme Heat on Economic Growth." *Science Advances,* Vol. 8, No. 43 (2022). https://www.science.org/doi/10.1126/sciadv.add3726

test scores: Christopher Flavelle. "Hotter Days Widen Racial Gap in U.S. Schools, Data Shows." *New York Times,* Oct. 5, 2020. https://www.nytimes.com/2020/10/05/climate/heat-minority-school-performance.html

miscarriage: Bruce Bekkar et al. "Association of Air Pollution and Heat Exposure With Preterm Birth, Low Birth Weight, and Stillbirth in the US: A Systematic Review." *JAMA Network Open* 3, no. 6 (2020). https://doi.org/10.1001/jamanetworkopen.2020.8243

heart and kidney disease: Barrak Alahmad et al. "Associations Between Extreme Temperatures and Cardiovascular Cause-Specific Mortality: Results From 27 Countries." *Circulation* vol. 147, issue 1 (2023), 35–46. https://www.doi.org/10.1161/CIRCULATIONAHA.122.061832; Woo-Seok Lee et al. "High Temperatures and Kidney Disease Morbidity: A Systematic Review and Meta-analysis."

Journal of Preventative Medicine & Public Health 52, vol. 1 (2019), 1–13. https://doi.org/10.3961%2Fjpmph.18.149

more impulsive: Yoonhee Kim et al. "Suicide and Ambient Temperature: A Multi-Country Multi-City Study." *Environmental Health Perspectives* 127, vol. 11 (2019). https://doi.org/10.1289/EHP4898

prone to conflict: Andreas Miles-Novelo and Craig A. Anderson. "Climate Change and Psychology: Effects of Rapid Global Warming on Violence and Aggression." *Current Climate Change Reports* 5 (2019), 36–46. https://doi.org/10.1007/s40641-019-00121-2

Racial slurs: Annika Stechemesser et al. "Temperature Impacts on Hate Speech Online: Evidence from 4 Billion Geolocated Tweets from the USA." *The Lancet Planetary Health* 6, no. 9 (2022), 714–725. https://doi.org/10.1016/S2542-5196(22)00173-5

Suicides rise: Kim et al.

Gun violence increases: Damian Carrington. "Almost 8,000 US Shootings Attributed to Unseasonable Heat." *The Guardian,* December 16, 2022. https://www.theguardian.com/world/2022/dec/16/almost-8000-us-shootings-attributed-to-unseasonable-heat-study

more rapes: Josephus Daniel Perry and Miles

E. Simpson. "Violent Crimes in a City: Environmental Determinants." *Environment and Behavior* 19, no. 1 (1987), 77–90. https://doi.org/10.1177/0013916587191004

civil war: Marshall B. Burke et al. "Warming Increases the Risk of Civil War in Africa." *Proceedings of National Academy of Sciences* 106, no. 49 (2009), 20670–20674. https://doi.org/10.1073/pnas.0907998106

356,000 people: Katrin G Burkart et al. "Estimating the Cause-Specific Relative Risks of Non-Optimal Temperature on Daily Mortality: a Two-Part Modelling Approach Applied to the Global Burden of Disease Study." *The Lancet* 398, no. 10301 (2021), 685–697. https://doi.org/10.1016/S0140-6736(21)01700-1

260,000: Rebecca R. Buchholz et al. "New Seasonal Pattern of Pollution Emerges from Changing North American Wildfires." *Nature Communications* 13, no. 2043 (2022). https://doi.org/10.1038/s41467-022-29623-8

spikes in hospitalizations: Megan Sever. "Western Wildfires' Health Risks Extend Across the Country." *ScienceNews,* June 17, 2022. https://www.sciencenews.org/article/wildfire-health-risks-air-smoke-west-east-united-states

2,000 feet of ice: William J. Broad. "How

the Ice Age Shaped New York." *New York Times,* June 5, 2018. https://www.nytimes.com/2018/06/05/science/how-the-ice-age-shaped-new-york.html

125,000 years ago: Andrew Dessler. *Introduction to Modern Climate Change* (Cambridge, England: Cambridge University Press, 2022), 33.

forty-year period: "Global Warming of 1.5°C." Intergovernmental Panel on Climate Change Special Report, 2018. https://www.ipcc.ch/sr15/

The eight years between 2015 and 2022: Robert Rohde. "Global Temperature Report for 2022." Berkeley Earth website. January 12, 2023. https://berkeleyearth.org/global-temperature-report-for-2022/

850 million people: Ibid.

longer, hotter: *Attribution of Extreme Weather Events in the Context of Climate Change* (Washington, DC: National Academies Press, 2016), 91.

150 times: "Western North American Extreme Heat Virtually Impossible Without Human-Caused Climate Change." *World Weather Attribution,* July 7, 2021. https://www.worldweatherattribution.org/western-north-american-extreme-heat-virtually-impossible-without-human-caused-climate-change/

The ocean: Damian Carrington. "Oceans Were Hottest Ever Recorded in 2022, Analysis Shows." *The Guardian,* January 11, 2023. https://www.theguardian.com/environment/2023/jan/11/oceans-were-the-hottest-ever-recorded-in-2022-analysis-shows

70 degrees: Jason Samenow and Kasha Patel. "It's 70 Degrees Warmer Than Normal in Eastern Antarctica. Scientists are Flabbergasted." *Washington Post,* March 18, 2022. https://www.washingtonpost.com/weather/2022/03/18/antarctica-heat-wave-climate-change/

three billion people: Chi Xu et al. "Future of the Human Climate Niche." *Proceedings of the National Academy of Sciences* 117, no. 21 (2020), 11350–11355. https://doi.org/10.1073/pnas.1910114117

life-threatening combinations: Camilo Mora et al. "Global Risk of Deadly Heat." *Nature Climate Change* 7 (2017), 501–506. https://doi.org/10.1038/nclimate3322

"fittest of humans": Eun-Soon Im, Jeremy S. Pal, and Elfatih A. B. Eltahir. "Deadly Heat Waves Projected in the Densely Populated Agricultural Regions of South Asia." *Science Advances* vol. 3, issue 8 (2017). https://doi.org/10.1126/sciadv.1603322

"Goldilocks Zone": Andrew May. "The Goldilocks Zone: The Place in a Solar System that's Just Right." *Live Science,* April 1, 2022. https://www.livescience.com /goldilocks-zone

Chapter 1 — A Cautionary Tale

"very active family": Mariposa County Sheriff's Report, case number MG2100896. August 18, 2021. Supplement 01, 1.

"Hi, please . . .": Facebook post accessed October 2022. https://www.facebook.com /sjeffe

"disturbed dirt": Mariposa County Sheriff's Report, August 19, 2021. Supplement 06, 2.

"they fell in love": Steve Rubenstein. "Remote Hiking Area Where Northern California Family Was Found Dead Treated as a Hazmat Site." *San Francisco Chronicle,* August 18, 2021. https://www.sfchronicle. com/bayarea/article/Remote-hiking-area -where-Northern-California-16395803.php

Richard wrote: "Down and Out: A Collection of Tales from My 20 Years As a Cave Explorer." Self-published on Richard Gerrish's website. Accessed October 2022. https://richardgerrish.weebly.com/down -and-out.html

"happier than Jonny": Personal communication with the author, September 2022.

"too stuffy": Mariposa County Sheriff's Report. Supplement 10, 1.

sixteen hikes: AllTrails post, accessed July 2022.

"city folk": Mariposa County Sheriff's Report. Supplement 10, 1.

2018 Ferguson Fire: Jose A. Del Rio. "Ferguson Fire Forces Largest Closing of Yosemite in Decades." *New York Times,* July 26, 2018. https://www.nytimes.com/2018/07/26/us/california-today-ferguson-fire-yosemite.html

"horrible trail": AllTrails post, accessed July 2022.

"hot day": Ibid.

95 degrees: Steven C. Sherwood and Matthew Huber. "An Adaptability Limit to Climate Change Due to Heat Stress." *Proceedings of the National Academy of Sciences* 107, no. 21 (2010), 9552–9555. https://doi.org/10.1073/pnas.0913352107

Kelly Watt: Amby Burfoot. "The Last Run." *Runner's World,* January 18, 2007. https://www.runnersworld.com/runners-stories/a20801399/the-dangers-of-running-in-the-heat/

Philip Kreycik: Sarah Trent. "Philip Kreycik Wasn't Supposed to Die This Way."

Outside, May 27, 2022. https://www.outsideonline.com/outdoor-adventure/environment/heat-related-illness-trail-running-death-philip-kreycik/

Michael Popov: Gordon Wright. "Michael Popov's Last Run." *Outside,* August 15, 2012. https://www.outsideonline.com/health/running/michael-popovs-last-run-coming-grips-sudden-death-exceptional-ultrarunner/#close

one study in Montana: John S. Cuddy and Brent S. Rudy. "High Work Output Combined with High Ambient Temperatures Caused Heat Exhaustion in a Wildland Firefighter Despite High Fluid Intake." *Wilderness & Environmental Medicine* 22, no. 2 (2011), 122–125. https://doi.org/10.1016/j.wem.2011.01.008

"Both heat exhaustion and heatstroke": Personal communication with the author, April 2022.

"exercise hard enough": Ibid.

heart and kidney disease: Cecilia Sorensen, MD, and Ramon Garcia-Trabanino, MD. "A New Era of Climate Medicine — Addressing Heat Triggered Renal Disease." *New England Journal of Medicine* 381, no. 8 (2019), 693–696. https://doi.org/10.1056/NEJMp1907859

Pregnant women: Raymond Zhong. "How

Extreme Heat Kills, Sickens, Strains and Ages Us." *New York Times,* June 13, 2022. https://www.nytimes.com/2022/06/13/climate/extreme-heat-wave-health.html

send a text: Joshua Bote. "'Can you help us': Final Text from California Family Found Dead on Hike Near Yosemite Released." *SFGate,* February 18, 2022. https://www.sfgate.com/bayarea/article/Final-texts-released-Chung-Gerrish-deaths-16930376.php

twenty-seven minutes: Ibid.

"**the physiological changes"** Nigel A. S. Taylor. "Human Heat Adaptation." *Comprehensive Physiology* 4 (2014), 325–365. https://doi.org/10.1002/cphy.c130022

"**if you go out of the heat":** Personal communication with the author, April 2022.

"**denature":** Veronique Greenwood. "Why Does Heat Kill Cells?" *The Atlantic,* May 11, 2017. https://www.theatlantic.com/science/archive/2017/05/heat-kills-cells/526377/

no obvious cause: "Federal Officials Close River After Mysterious Deaths of California Family and Their Dog." *CBS News,* September 6, 2021. https://www.cbsnews.com/news/john-gerrish-ellen-chung-daughter-deaths-merced-river-closed/

"**carbon monoxide":** Michelle Blade. "Carbon Monoxide Could Have Killed Lancaster

Man and His Family on California Hiking Trail." *Lancaster Guardian,* August 19, 2021. https://www.lancasterguardian.co.uk/news/people/carbon-monoxide-could-have-killed-lancaster-man-and-his-family-on-california-hiking-trail-3352070

Anatoxin A: Mariposa County Sheriff's Report, August 26, 2021. Supplement 16, 1.

"death like this": Adrian Thomas. "Mariposa County Sheriff: 'I've Never Seen a Death Like This.'" Yourcentralvalley.com, August 18, 2021. https://www.yourcentralvalley.com/news/local-news/mariposa-county-sheriff-ive-never-seen-a-death-like-this/

dogs: Emily J. Hall et al. "Incidence and Risk Factors for Heat-Related Illness (Heatstroke) in UK Dogs Under Primary Veterinary Care in 2016." *Scientific Reports* 10, no. 9128 (2020). https://doi.org/10.1038/s41598-020-66015-8

"Greyhounds": Eric Roston. "These Very Good Dogs Will Suffer Most from a Warming Climate." *Bloomberg Green,* June 19, 2020. https://www.bloomberg.com/news/photo-essays/2020-06-19/these-dog-breeds-are-the-most-vulnerable-to-heat-climate-change

prepubescent children: Leigh Arlegui

et al. “Body Mapping of Sweating Patterns of Pre-Pubertal Children During Intermittent Exercise in a Warm Environment.” *European Journal of Applied Physiology* 121 (2021). https://hdl.handle.net/2134/16831309.v1

“dad-bod”: Personal communication with the author, April 2022.

luteal phase: Nisha Charkoudian and Nina S. Stachenfeld. “Reproductive Hormone Influences on Thermoregulation in Women.” *Comprehensive Physiology* 4, no. 2 (2014). https://doi.org/10.1002/cphy.c130029

birth control pills: Nisha Charkoudian et al. “Autonomic Control of Body Temperature and Blood Pressure: Influences of Female Sex Hormones.” *Clinical Autonomic Research* 27 (2017). https://doi.org/10.1007/s10286-017-0420-z

“The cause of death”: Mariposa County Sheriff’s Report, October 19, 2021. Supplement 50, 1.

Chapter 2 — How Heat Shaped Us

widely accepted theory: Carl Zimmer. *Life’s Edge: The Search for What It Means to Be Alive* (New York: Dutton, 2021), 242–246.

“reached the sea”: Ibid., 245.

260 million years ago: Kévin Rey et al. "Oxygen Isotopes Suggest Elevated Thermometabolism Within Multiple Permo-Triassic Therapsid Clades." *eLife* 6 (2017). https://doi.org/10.7554/eLife.28589

thirty times less energy: Michael Logan. "Did Pathogens Facilitate the Rise of Endothermy?" *Ideas in Ecology and Evolution* 12 (2019), 1–8. https://doi.org/10.24908/iee.2019.12.1.n.

"Imagine an iguana": Personal communication with the author, April 2021.

Lucy: Lewis Dartnell. *Origins: How Earth's History Shaped Human History* (New York: Basic Books, 2019), 14–15.

an evolutionary toddler: Other archaeological finds push the date for bipedalism back even further. The Laetoli footprints, found in Tanzania by Louis Leakey in 1976, date to 3.7 million years ago. In 1994, *Ardipithecus ramidus* fossils were found with features that suggest bipedalism might have originated more than 4.4 million years ago. See also Clare Wilson, "Human Ancestors May Have Walked on Two Legs 7 Million Years Ago," *New Scientist,* August 24, 2022.

what made Lucy stand up: Daniel E. Liberman. "Human Locomotion and Heat Loss: An Evolutionary Perspective."

Comprehensive Physiology 5 (2015), 99–117. https://doi.org/10.1002/cphy.c140011

Elephants: Robin C. Dunkin et al. "Climate Influences Thermal Balance and Water Use in African and Asian Elephants: Physiology Can Predict Drivers of Elephant Distribution." *Journal of Experimental Biology* 216, no. 15 (2013), 2939–2952. https://doi.org/10.1242/jeb.080218

silver Saharan ant: Jake Buehler. "World's Fastest Ants Found Racing Across the Sahara." *National Geographic,* October 16, 2019. https://www.nationalgeographic.com/animals/article/silver-saharan-ants-fastest-desert

camels: Mulu Gebreselassie Gebreyohanes and Awol Mohammed Assen. "Adaptation Mechanisms of Camels (Camelus dromedarius) for Desert Environment: A Review." *Journal of Veterinary Science & Technology* 8, no. 6 (2017). https://doi.org/10.4172/2157-7579.1000486

Fongoli chimps: Carl Zimmer. "Hints of Human Evolution in Chimpanzees That Endure a Savanna's Heat." *New York Times,* April 27, 2018. https://www.nytimes.com/2018/04/27/science/chimpanzees-savanna-evolution.html

The Lion King: Dartnell, *Origins,* 12.

bipedalism: Kevin Hunt. *Chimpanzee:*

Lessons from Our Sister Species (Cambridge, England: Cambridge University Press, 2021), 480–484.

a taste for spicy foods: Rick Weiss. "Healthy Hypothesis Curries Favor in Evolution of Spice." *Washington Post,* March 2, 1998. https://www.washingtonpost.com/archive/politics/1998/03/02/health-hypothesis-curries-favor-in-evolution-of-spice/e025c5ab-76c4-4d71-b1bc-9b36cef22bc0/

Chapter 3 — Heat Islands

Flights are delayed: "Phoenix Flights Cancelled Because It's Too Hot for Planes." *BBC News,* June 20, 2017. https://www.bbc.com/news/world-us-canada-40339730

twenty-one degrees hotter: Personal communication with Phoenix office of National Weather Service, May 2022.

twenty degrees warmer: "Cool Neighborhoods NYC: A Comprehensive Approach to Keep Cities Cool in Extreme Heat." Report by the City of New York, Office of the Mayor. https://www1.nyc.gov/assets/orr/pdf/Cool_Neighborhoods_NYC_Report.pdf

339 heat-related deaths: Maricopa County Public Health. "Heat-Associated Deaths in Maricopa County, AZ Final Report for 2021." https://www.maricopa.gov/ArchiveCenter/ViewFile/Item/5494

70,000 people: Friederike Otto. *Angry Weather: Heat Waves, Floods, Storms and the New Science of Climate Change* (Berkeley/Vancouver: Greystone, 2020), 94.

1.7 billion people: Cascade Tuholske et al. "Global Urban Population Exposure to Extreme Heat." *Proceedings of the National Academy of Sciences* 118 (2021). https://doi.org/10.1073/pnas.2024792118

70 percent: "68% of the World Population Projected to Live in Urban Areas by 2050, Says UN." United Nations website, May 16, 2018. Accessed October 2022. https://www.un.org/development/desa/en/news/population/2018-revision-of-world-urbanization-prospects.html

Hurricane Katrina: National Weather Service. "Extremely Powerful Hurricane Katrina Leaves a Historic Mark on the Northern Gulf Coast." https://www.weather.gov/mob/katrina

a utility worker: Marty Graham. "Power Restored in Southwest, Mexico After Outage." *Reuters,* September 8, 2011. https://www.reuters.com/article/us-outage-california/power-restored-in-southwest-mexico-after-outage-idUSTRE7880FW20110909

Frederic Tudor: A. S. Ganesh. "The Ice King of the Past." *The Hindu,* September 13, 2020. https://www.thehindu.com/children/the-ice-king-of-the-past/article32529190.ece

100 square miles: "How One of the World's Wettest Major Cities Ran Out of Water." *Bloomberg News,* February 3, 2021. https://www.bloomberg.com/news/features/2021-02-03/how-a-water-crisis-hit-india-s-chennai-one-of-the-world-s-wettest-cities

water into the city: Ibid.

123 degrees: Mujib Mashal. "India Heat Wave, Soaring Up to 123 Degrees, Has Killed at Least 36." *New York Times,* June 13, 2019. https://www.nytimes.com/2019/06/13/world/asia/india-heat-wave-deaths.html

poor water storage: "How One of the World's Wettest Major Cities Ran Out of Water."

one journalist wrote: Ibid.

"safer" housing: R. K. Radhakrishnan. "Flood of Troubles." *Frontline,* May 27, 2016. https://frontline.thehindu.com/the-nation/flood-of-troubles/article8581086.ece

Stephanie Pullman: Elizabeth Whitman. "On 107-Degree Day, APS Cut

Power to Stephanie Pullman's Home. She Didn't Live." *Phoenix New Times,* June 3, 2019. https://www.phoenixnewtimes.com /content/printView/11310515

Phoenix New Times: Ibid.

110,000 times: Ibid.

39,000 cutoffs: Ibid.

Chapter 4 — Life on the Run

the "Mother Road": Rickie Longfellow. "Route 66: the Mother Road." US Department of Transportation website. Accessed October 2022. https://www.fhwa.dot.gov /infrastructure/back0303.cfm

"We loaded up our jalopies": Woody Guthrie. "The Great Dust Storm." https://www .woodyguthrie.org/Lyrics/Dust_Storm _Disaster.htm

40 and 70 percent: Sonia Shah. *The Next Great Migration: The Beauty and Terror of Life on the Move* (New York: Bloomsbury, 2020), 5.

twenty kilometers: Ibid.

four times faster: Ibid.

Atlantic cod: Ibid.

frogs and fungi: Ibid.

coral polyps: Ibid.

two miles per decade: James Bridle. "The Speed of a Tree: How Plants Migrate to Outpace Climate Change."

Financial Times, April 1, 2022. https://www.ft.com/content/7d7621cd-7bb5-4f97-94f1-6985ce038e13

white spruce: Ibid.

Thick-billed murres: Emily S. Choy et al. "Limited Heat Tolerance in a Cold-Adapted Seabird: Implications of a Warming Arctic." *Journal of Experimental Biology* 224, no. 13 (2021). https://doi.org/10.1242/jeb.242168

"migration barrier": Matthew L. Keefer et al. "Thermal Exposure of Adult Chinook Salmon and Steelhead: Diverse Behavioral Strategies in a Large and Warming River System." *PLoS One* 13, no. 9 (2018). https://doi.org/10.1371/journal.pone.0204274

taxied young salmon: Christian Martinez. "Wildlife Officials Truck Chinook Salmon to Cooler Waters in Emergency Move to Help Them Spawn." *Los Angeles Times,* May 20, 2022. https://www.latimes.com/california/story/2022-05-19/northern-california-chinook-salmon-trucked-to-cooler-waters

"Bumblebees are disappearing": Quoted in Bob Brewyn, "Bumblebee Decline Linked With Extreme Heat Waves." *Inside Climate News,* February 6, 2020. https://insideclimatenews.org/news/06022020/bumblebee-climate-change-heat-decline-migration/

ten times larger: Personal communication with the author, October 2022.

highest population growth rates: "More Than Half of U.S. Counties Were Smaller in 2020 Than in 2010." US Census Bureau website. Accessed October 2022. https://www.census.gov/library/stories/2021/08/more-than-half-of-united-states-counties-were-smaller-in-2020-than-in-2010.html

the Dust Bowl: Richard Hornbeck. "The Enduring Impact of the American Dust Bowl: Short- and Long-Run Adjustments to Environmental Catastrophe." *American Economic Review,* American Economic Association 102, no. 4 (2012), 1477–1507. https://www.nber.org/papers/w15605

eight million people: World Bank Group. "Groundswell: Preparing for Internal Climate Migration." World Bank report, 2018. https://www.worldbank.org/en/news/infographic/2018/03/19/groundswell —preparing-for-internal-climate-migration

"Should the flight away": Abrahm Lustgarten. "The Great Climate Migration." *New York Times,* August 23, 2020. https://www.nytimes.com/interactive/2020/07/23/magazine/climate-migration.html

floods in Pakistan: Christina Goldbaum and Zia ur-Rehman. "In Pakistan's Record Floods, Villages Are Now

Desperate Islands." *New York Times,* September 14, 2022. https://www.nytimes.com/2022/09/14/world/asia/pakistan-floods.html

the fifty US counties: Lily Katz and Sebastian Sandoval-Olascoaga. "More People Are Moving In Than Out of Areas Facing High Risk From Climate Change." *Redfin News,* August 25, 2021. https://www.redfin.com/news/climate-migration-real-estate-2021/

a 2020 report: US Government Accountability Office. "Climate Change: A Climate Migration Pilot Program Could Enhance the Nation's Resilience and Reduce Federal Fiscal Exposure." GAO Report to Congressional Requesters, July 2020. https://www.gao.gov/assets/gao-20-488.pdf

storms: Katz and Sandoval-Olascoaga, "More People Are Moving In Than Out."

the most attractive hot place: Ibid.

"If you live in Kansas": Quoted in Saul Elbein, "Five Reasons Extreme Weather is Bigger in Texas." *The Hill,* September 1, 2022. https://thehill.com/policy/equilibrium-sustainability/3622655-five-reasons-extreme-weather-is-bigger-in-texas/

a previous book I wrote: Jeff Goodell. *The*

Water Will Come: Rising Seas, Sinking Cities, and the Remaking of the Civilized World (New York: Little, Brown, 2017.)

freakish ice storm: 246 Texans died during the storm (the actual toll is likely far higher), many of them freezing to death in their homes or in the streets. Patrick Svitek. "Texas Puts Final Estimate of Winter Storm Death Toll at 246." *Texas Tribune,* January 2, 2022. https://www.texastribune.org/2022/01/02/texas-winter-storm-final-death-toll-246/

a haunted place: Ghosts of lost travelers are everywhere in Luis Alberto Urrea's *The Devil's Highway: A True Story* (New York: Little, Brown, 2004).

nine thousand people: Jason Motlagh. "The Deadliest Crossing." *Rolling Stone,* September 30, 2019. https://www.rollingstone.com/politics/politics-features/border-crisis-arizona-sonoran-desert-882613/

Gurupreet Kaur: Sugam Pokharel and Catherine E. Shoichet. "This 6-Year-Old From India Died in the Arizona Desert. She Loved Dancing and Dreamed of Meeting Her Dad." *CNN,* July 12, 2019. https://www.cnn.com/2019/07/12/asia/us-border-death-indian-girl-family/index.html

Chapter 5 — Anatomy of a Crime Scene

"Physicist, wannabe-dancer . . .": Simon Levey. "Climate Scientist in TIME100 Most Influential List to Join Imperial." Imperial College *London News,* September 15, 2021. https://www.imperial.ac.uk/news/229993/climate-scientist-time100-most-influential-list/

Egyptians: Joshua J. Mark. "Ancient Egyptian Warfare." *World History Encyclopedia*. https://www.worldhistory.org/Egyptian_Warfare/

Hinduism: "tapas," *Encyclopedia Britannica,* February 28, 2011. https://www.britannica.com/topic/tapas.

Native Americans: Joseph Bruchag. *The Native American Sweat Lodge: History and Legends* (Freedom, CA: Crossing Press, 1993), 24–29.

"heat and fire": Martin Goldstein and Inge F. Goldstein. *The Refrigerator and the Universe: Understanding the Laws of Energy* (Cambridge, MA: Harvard University Press, 1993), 29.

"heat occurs here": *Ibn-Sina–Al-Biruni Correspondence* (Alberta, Canada: Center for Islam and Science, 2003), 8.

"detached rays of the sun": Razaullah Ansari. "On the Physical Researches of Al-Biruni." *International Journal of Health*

Sciences 10, no. 2 (1975), 198–217.

Galileo: Martin Goldstein and Inge F. Goldstein, *The Refrigerator and the Universe,* 33–34.

Fahrenheit: Phil Jaekl. "Melting Butter, Poisonous Mushrooms and the Strange History of the Invention of the Thermometer." *Time,* June 1, 2021. https://time.com/6053214/thermometer-history/

Count Rumford: In recounting Count Rumford's life and work, I drew from G. I. Brown, *Scientist, Soldier, Statesman, Spy: Count Rumford, the Extraordinary Life of a Scientific Genius* (United Kingdom: Sutton Publishing, 1999), and Sanborn C. Brown, *Benjamin Thompson, Count Rumford* (Cambridge, MA: MIT Press, 1981). Also helpful was Hans Christian Von Baeyer, *Warmth Disperses and Time Passes: The History of Heat* (New York: Modern Library, 1999).

giddy biographer: Jane Merrill. *Sex and the Scientist: The Indecent Life of Benjamin Thompson, Count Rumford (1753–1814)* (Jefferson, North Carolina: McFarland & Company, 2018).

"trembling agitation": D. S. L. Cardwell. *From Watt to Clausius: The Rise of Thermodynamics in the Early Industrial Age* (Ithaca, NY: Cornell University Press, 1971), 33.

caloric theory: Martin Goldstein and Inge F. Goldstein, *The Refrigerator and the Universe,* 29–35.

"kill Frenchmen": Von Baeyer, *Warmth Disperses and Time Passes,* 3.

"It would be difficult to describe": Brown, *Scientist, Soldier, Statesman, Spy,* 86.

"laughter through a crowd": Brian Greene. *Until the End of Time: Mind, Matter, and Our Search for Meaning in an Evolving Universe* (New York: Knopf, 2020), 87.

"equable and better climates": Fred Pearce. "Land of the Midnight Sums." *New Scientist,* January 25, 2003. https://www.newscientist.com/article/mg17723795-300-land-of-the-midnight-sums/

"Nobody worried about the impacts": Spencer R. Weart. *The Discovery of Global Warming* (Cambridge, MA: Harvard University Press, 2008). Accessed online October 2022. https://history.aip.org/climate/impacts.htm

"a violent effect": Ibid.

"real deserts": Revelle in United States Congress, House of Representatives, Committee on Appropriations, *Report on the International Geophysical Year* (Washington, DC: Government Printing Office, 1957), 104–106.

1988 Congressional testimony: Philip Shabecoff. "Global Warming Has Begun, Expert Tells Senate." *New York Times,* June 24, 1988.

seventy thousand people: Otto, *Angry Weather,* 94.

Allen's commentary: Myles Allen. "Liability for Climate Change." *Nature* 421 (2003), 891–892. https://doi.org/10.1038/421891a

"doubled the risk": Peter Stott et al. "Human Contribution to the European Heatwave of 2003." *Nature* 432 (2004), 610–614. https://doi.org/10.1038/nature03089

55,000 people: Hannah Hoag. "Russian Summer Tops 'Universal' Heatwave Index." *Nature,* October 29, 2014. https://doi.org/10.1038/nature.2014.16250

"internal atmospheric variability": Otto, *Angry Weather,* 64.

80 percent probability: Ibid.

"install electric lanterns": Ibid, 84.

"three to four times more likely": Ibid, 105.

Bangladesh: Sjoukje Philip et al. "Attributing the 2017 Bangladesh floods from Meteorological and Hydrological Perspectives." *Hydrology and Earth System Sciences* 23 (2019), 1409–1429. https://doi.org/10.5194/hess-23-1409-2019

"virtually impossible": Sjoukje Philip

et al. "Rapid Attribution Analysis of the Extraordinary Heatwave on the Pacific Coast of the US and Canada June 2021." Self-published by World Weather Attribution. https://www.worldweatherattribution.org/wp-content/uploads/NW-US-extreme-heat-2021-scientific-report-WWA.pdf

"heat waves in Europe": Efi Rousi et al. "Accelerated Western European Heatwave Trends Linked to More-Persistent Double Jets Over Eurasia." *Nature Communications* 13, no. 3851 (2022). https://doi.org/10.1038/s41467-022-31432-y

"Everybody is really worried": Chelsea Harvey. "Heat Wave 'Virtually Impossible' Without Climate Change." *E&E News,* July 8, 2021. https://www.eenews.net/articles/heat-wave-virtually-impossible-without-climate-change/

ExxonMobil: "Update of Carbon Majors 1965–2018." Climate Accountability Institute, December 20, 2020. https://climateaccountability.org/pdf/CAI%20PressRelease%20Dec20.pdf

Chapter 6 — Magic Valley

"crunchy, dry, and dusty": Monique Brand. "As Temperatures Rise, Agriculture Industry Suffers." *Lampasas*

Dispatch Record, July 25, 2022. https://www.lampasasdispatchrecord.com/news/temperatures-rise-agriculture-industry-suffers

42 percent of Texas corn acreage: Bob Sechler. "'Difficult Times': Heat, Drought Bringing Pain for Texas Farmers and Ranchers." *Austin American-Statesman,* July 25, 2022. https://www.statesman.com/story/business/economy/2022/07/25/texas-weather-heat-drought-farmers-ranchers-impact-bringing-pain/65377249007/

"no corn to sell": Ibid.

three decades: Megan Durisin. "Smallest French Corn Crop Since 1990 Shows Drought's Huge Toll." *Bloomberg,* September 13, 2022. https://www.bloomberg.com/news/articles/2022-09-13/smallest-french-corn-crop-since-1990-shows-drought-s-huge-toll

"Look at what the heat wave did": Arshad R. Zargar. "Wheat Prices Hit Record High as India's Heat Wave-driven Export Ban Compounds Ukraine War Supply Woes." *CBS News,* May 17, 2022. https://www.cbsnews.com/news/india-heat-wave-wheat-prices-soar-climate-change-ukraine-war-supplies/

20 million tons: Wailin Wong. "Russia Has Blocked 20 Million Tons of Grain from Being Exported from Ukraine." *All Things Considered,* June 3, 2022. https://www.npr.org/2022/06/03/1102990029/russia-has-blocked-20-million-tons-of-grain-from-being-exported-from-ukraine

60 percent: Ibid.

additional twenty-three million people: Eyder Peralta. "Drought and Soaring Food Prices from Ukraine War Leave Millions in Africa Starving." *NPR,* May 18, 2022. https://www.npr.org/2022/05/18/1099733752/famine-africa-ukraine-invasion-drought

sparked riots: Jen Kirby. "Sri Lanka's protests are just the beginning of global instability." *Vox,* July 16, 2022. https://www.vox.com/23211533/sri-lanka-protests-food-fuel-ukraine-war

"**By attacking Ukraine":** Edward Wong and Ana Swanson. "How Russia's War on Ukraine Is Worsening Global Starvation" *New York Times,* January 2, 2023. https://www.nytimes.com/2023/01/02/us/politics/russia-ukraine-food-crisis.html

345 million: Yuka Hayashi. "Ukraine War Creates Worst Global Food Crisis Since 2008, IMF Says." *Wall Street Journal,* September 20, 2022. https://www.wsj.com/articles/ukraine-war-creates-worst-global-food-crisis-since-2008-imf-says-11664553601

wasted: "Food Loss and Waste." US Food & Drug Administration website. Accessed October 2022. https://www.fda.gov/food/consumers/food-loss-and-waste
30 million acres: Michael Grunwald. "Biofuels Are Accelerating the Food Crisis — and the Climate Crisis, Too." *Canary Media,* April 19, 2022. https://www.canarymedia.com/articles/food-and-farms/biofuels-are-accelerating-the-food-crisis-and-the-climate-crisis-too
three feet a year: Elizabeth Kolbert. "Creating a Better Leaf." *The New Yorker,* December 6, 2021. https://www.newyorker.com/magazine/2021/12/13/creating-a-better-leaf
World Resources Institute: Tim Searchinger et al. "Creating a Sustainable Food Future." World Resources Report, 2019. https://research.wri.org/wrr-food
21 percent lower: Ariel Ortiz-Bobea et al. "Anthropogenic Climate Change Has Slowed Global Agricultural Productivity Growth." *Nature Climate Change* 11 (2021), 306–312. https://doi.org/10.1038/s41558-021-01000-1
For every degree Celsius of increase: Chuang Zhao et al. "Temperature Increase Reduces Global Yields of Major Crops in Four Independent Estimates."

Proceedings of the National Academy of Sciences 114, no. 35 (2017), 9326–9331. https://doi.org/10.1073/pnas.1701762114

"how important temperature is": David Lobell. "Heat and Hunger." Talk at Arizona State University, March 25, 2013. https://sustainability-innovation.asu.edu/media/video/david-lobell/

four thousand gallons a day: Eric J. Wallace. "Americans Have Planted So Much Corn That It's Changing the Weather." *Atlas Obscura,* December 3, 2018. https://www.atlasobscura.com/articles/corn-belt-weather

dry heat much more damaging: Mingfang Ting et al. "Contrasting Impacts of Dry Versus Humid Heat on US Corn and Soybean Yields" *Scientific Reports* 13, article 710 (2023). https://doi.org/10.1038/s41598-023-27931-7

arsenic: E. Marie Muehe et al. "Rice Production Threatened by Coupled Stresses of Climate and Soil Arsenic." *Nature Communications* 10, no. 4985 (2019). https://doi.org/10.1038/s41467-019-12946-4

breast and bladder cancer: "Rice Consumption and Cancer Risk." Columbia Public Health. Accessed October 2022. https://www.publichealth.columbia.edu/research/niehs-center-environmental

-health-northern-manhattan/rice-consumption-and-cancer-risk

high-CO_2 conditions: Chunwu Zhu et al. "Carbon Dioxide (CO_2) Levels This Century Will Alter the Protein, Micronutrients, and Vitamin Content of Rice Grains with Potential Health Consequences for the Poorest Rice-Dependent Countries." *Science Advances* 4, no. 5 (2018). https://doi.org/10.1126/sciadv.aaq1012

Magic Valley: Naveena Sadasivam. "The Making of the 'Magic Valley.'" *Texas Observer,* August 21, 2018. https://www.texasobserver.org/the-making-of-the-magic-valley/

twelve hundred years: A. Park Williams et al. "Rapid Intensification of the Emerging Southwestern North American Megadrought in 2020–2021." *Nature Climate Change* 12 (2022), 232–234. https://doi.org/10.1038/s41558-022-01290-z

"extreme drought": US Drought Monitor. Accessed February 2022. https://droughtmonitor.unl.edu

"It's like a crime scene": Danielle Prokop. "It's Like a Crime Scene What's Happened to the Rio Grande in Far West Texas." *Source New Mexico,* February 22, 2023. https://sourcenm.com/2023/02/22/its-like-a-crime-scene-whats-happened

-to-the-rio-grande-in-far-west-texas/

hold its breath: University of California Davis Botanical Conservatory. "The Genus *Aloe.*" *Botanical Notes* 1, no. 1 (July 2009). https://greenhouse.ucdavis.edu/files/botnot_01-01.00.pdf

temporary hibernation: Ibid.

the freeze that hit Texas: Jeff Goodell. "Is Texas' Disaster a Harbinger of America's Future?" *Rolling Stone,* February 17, 2021. https://www.rollingstone.com/culture/culture-news/austin-texas-ice-storm-climate-change-1129183/

Balsas River Valley: Anthony J. Ranere et al. "The Cultural and Chronological Context of Early Holocene Maize and Squash Domestication in the Central Balsas River Valley, Mexico." *Proceedings of the National Academy of Sciences* 106, no. 13 (2009), 5014–5018. https://doi.org/10.1073/pnas.0812590106

half the corn: Aaron Viner. "Ethanol Continues to Sustain Corn Prices." *Iowa Farmer Today,* May 26, 2022. https://www.agupdate.com/iowafarmertoday/news/state-and-regional/ethanol-continues-to-sustain-corn-prices/article_2ed66ffc-aabe-11ec-9e9c-cb70dac4ee17.html

French Revolution: Lisa Bramen. "When Food Changed History: the French

Revolution." *Smithsonian,* July 14, 2010. https://www.smithsonianmag.com/arts-culture/when-food-changed-history-the-french-revolution-93598442

1917 Russian Revolution: "Russian Revolution of 1917." *Encyclopedia Britannica.* Accessed October 2022. https://www.britannica.com/summary/Russian-Revolution

Arab Spring uprising: Joshua Keating. "A Revolution Marches on its Stomach." *Slate,* April 8, 2014. https://slate.com/technology/2014/04/food-riots-and-revolution-grain-prices-predict-political-instability.html

"potential new crop species": Quoted in John McCracken, "The Corn Belt Will Get Hotter. Farmers Will Have to Adapt," *Grist,* September 23, 2022. https://grist.org/agriculture/the-corn-belt-will-get-hotter-farmers-will-have-to-adapt/

Market cap of $500 million: AppHarvest (APPH) has been on a wild ride, with a market cap of about $3.4 billion shortly after the company went public in early 2021, to a low of $172 million in October 2022.

thousands of cows: "Heat Stress Blamed for Thousands of Cattle Deaths in Kansas." *PBS News Hour,* June 17, 2022.

https://www.pbs.org/newshour/economy/heat-stress-blamed-for-thousands-of-cattle-deaths-in-kansas

Texas cattle fever: Rob Williams. "Texas Cattle Fever Back with a Vengeance." *Texas A&M AgriLife Communications,* February 2, 2017. https://entomology.tamu.edu/2017/02/02/texas-cattle-fever-ticks-are-back-with-a-vengeance/

boiled alive: Latika Bourke. "'Boiled Alive': New Footage Shows Full Scale of Live Exports Horror." *Sydney Morning Herald,* May 5, 2018. https://www.smh.com.au/politics/federal/boiled-alive-new-footage-shows-full-scale-of-live-exports-horror-20180503-p4zd9q.html

Fungi-based proteins: "Everything You Need to Know about Fungi-based Proteins." Nature's Fynd website. Accessed October 2022. https://www.naturesfynd.com/blog/fungi-based-protein

cricket farms: Rebecca Zandbergen. "Massive Cricket-Processing Facility Comes Online in London, Ont." *CBC News,* July 1, 2022. https://www.cbc.ca/news/canada/london/cricket-farm-london-ontario-1.6506606

Chapter 7 — The Blob

a high-pressure ridge: "Looking Back at the Blob: Record Warming Drives

Unprecedented Ocean Change." NOAA Fisheries News, September 26, 2019. https://www.fisheries.noaa.gov/feature-story/looking-back-blob-record-warming-drives-unprecedented-ocean-change

the Blob: Ibid. See also the "Blob Tracker" maintained by California Current Integrated Assessment. https://www.integratedecosystemassessment.noaa.gov/regions/california-current/cc-projects-blobtracker

sea lion strandings: Jon Brooks. "How 'The Blob' Has Triggered Disaster for California Seals," KQED, November 23, 2015. https://www.kqed.org/science/373789/how-warmer-waters-have-led-to-emaciated-seals-on-california-beaches

Alaska cod fishery: "Alaska Cod Populations Plummeted During the Blob Heatwave." NOAA Fisheries News, November 8, 2019. https://www.fisheries.noaa.gov/feature-story/alaska-cod-populations-plummeted-during-blob-heatwave-new-study-aims-find-out-why

kelp forests: Meredith L. McPherson et al. "Large-Scale Shift in the Structure of a Kelp Forest Ecosystem Co-Occurs with an Epizootic and Marine Heatwave." *Communications in Biology* 4, no. 298 (2021). https://doi.org/10.1038/s42003-021-01827-6

a million seabirds: Adam Vaughan. "Marine

Heatwave Known as 'the Blob' Killed a Million US Seabirds." *New Scientist,* January 15, 2020. https://www.newscientist.com/article/2229980-marine-heatwave-known-as-the-blob-killed-a-million-us-seabirds

Camp Fire: Stella Chan and Joe Sterling. "Death Toll in Camp Fire Revised Down by one to 85." *CNN,* February 8, 2019. https://www.cnn.com/2019/02/08/us/camp-fire-deaths

icy asteroids: Conel M. O'D. Alexander. "The Origin of Inner Solar System Water." *Philosophical Transactions of the Royal Society* vol. 375, issue 2094 (2017). https://dx.doi.org/10.1098/rsta.2015.0384

Neil Shubin: Personal communication with the author, January 14, 2020.

Eighty percent: "How Much of the Ocean Have We Explored?" NOAA Ocean Facts. Accessed October 2022. https://oceanservice.noaa.gov/facts/exploration.html

large fish: Bob Holmes. "Ocean's Great Fish All But Gone." *New Scientist,* May 14, 2003. https://www.newscientist.com/article/dn3731-oceans-great-fish-all-but-gone/

more plastic than fish: Sarah Kaplan. "By 2050, There Will Be More Plastic Than Fish in the Ocean, Study Says." *Washington Post,* January 20, 2016. https://

www.washingtonpost.com/news/morning-mix/wp/2016/01/20/by-2050-there-will-be-more-plastic-than-fish-in-the-worlds-oceans-study-says/

seven warmest years: Oliver Milman. "Hottest Ocean Temperatures in History Recorded Last Year." *The Guardian,* January 11, 2022. https://www.theguardian.com/environment/2022/jan/11/oceans-hottest-temperatures-research-climate-crisis

the ocean hit its warmest temperature: Chris Mooney and Brady Dennis. "Oceans Surged to Another Record-High Temperature in 2022." *Washington Post,* January 11, 2023. https://www.washingtonpost.com/climate-environment/2023/01/11/ocean-heat-climate-change/

top mile or so of the ocean: Lijing Cheng et al. "Past and Future Ocean Warming." *Nature Reviews Earth & Environment* 3 (2022) 776–794. https://doi.org/10.1038/s43017-022-00345-1

a hundred microwave ovens: Lijing Cheng et al. "Upper Ocean Temperatures Hit Record High in 2020." *Advances in Atmospheric Science* 38 (2021), 523–530. https://doi.org/10.1007/s00376-021-0447-x

atmospheric rivers: National Oceanic and Atmospheric Association. "Atmospheric

Rivers: What Are They and How Does NOAA Study Them?" *NOAA Research News,* January 11, 2023. https://research.noaa.gov/article/ArtMID/587/ArticleID/2926/Atmospheric-Rivers-What-are-they-and-how-does-NOAA-study-them

"a direct consequence": Quoted in Chris Mooney and Brady Dennis, "Oceans Surged to Another Record-High Temperature in 2022."

62,000 workers: "The Economic Value of Alaska's Seafood Industry." Report by Alaska Seafood Marketing Institute, January 2022. https://www.alaskaseafood.org/wp-content/uploads/MRG_ASMI-Economic-Impacts-Report_final.pdf

1.8 million people: "Fisheries Economics of the United States Report, 2019." NOAA Fisheries. https://www.fisheries.noaa.gov/resource/document/fisheries-economics-united-states-report-2019

$255 billion: Ibid.

The report's summary: *IPCC Special Report on the Ocean and Cryosphere in a Changing Climate* (Cambridge, England: Cambridge University Press, 2022), 18. https://doi.org/10.1017/9781009157964.

The Mediterranean: Jon Henley. "Mediterranean Ecosystem Suffering 'Marine

Wildfire' as Temperatures Peak." *The Guardian,* July 29, 2022. https://www.theguardian.com/environment/2022/jul/29/mediterranean-ecosystem-suffering-marine-wildfire-as-temperatures-peak

"underwater wildfires": Ibid.

Tasmania: Darryl Fears. "On Land, Australia's Rising Heat is 'Apocalyptic.' In the Ocean, It's Worse." *Washington Post,* December 27, 2019. https://www.washingtonpost.com/graphics/2019/world/climate-environment/climate-change-tasmania/

Uruguayan coast: Chris Mooney and John Muyskens. "Dangerous New Hot Zones Are Spreading Around the World." *Washington Post,* September 11, 2019. https://www.washingtonpost.com/graphics/2019/national/climate-environment/climate-change-world/

Nations in the tropics: Kimberly L. Oremus et al. "Governance Challenges for Tropical Nations Losing Fish Species Due to Climate Change." *Nature Sustainability* 3 (2020), 277–280. https://doi.org/10.1038/s41893-020-0476-y

"losing a stock": Quoted in Harrison Tusoff, "Fleeing Fish." *The Current,* February 24, 2020. https://www.news.ucsb.edu/2020/019806/fleeing-fish

six bleaching events: Damien Cave.

"'Can't Cope': Australia's Great Barrier Reef Suffers 6th Mass Bleaching Event." *New York Times,* March 25, 2022. https://www.nytimes.com/2022/03/25/world/australia/great-barrier-reef-bleaching.html

93 percent: Terry P. Hughes et al. "Emergent Properties in the Responses of Tropical Corals to Recurrent Climate Extremes." *Current Biology* 31, no. 23 (2021), 5393–5399. https://doi.org/10.1016/j.cub.2021.10.046

the Reef 2050 Plan: Reef 2050 Long-Term Sustainability Plan 2021–2025, Commonwealth of Australia 2021. Commonwealth of Australia. Accessed October 2022. https://www.dcceew.gov.au/parks-heritage/great-barrier-reef/long-term-sustainability-plan

2021 editorial: Terry Hughes. "The Great Barrier Reef Actually Is 'in Danger.'" *The Hill,* July 26, 2021. https://thehill.com/opinion/energy-environment/564778-the-great-barrier-reef-actually-is-in-danger/

lathering corals: Nicola Jones. "Finding Bright Spots in the Global Coral Reef Catastrophe." *Yale e360,* October 21, 2021. https://e360.yale.edu/features/finding-bright-spots-in-the-global-coral-reef-catastrophe

Chapter 8 — The Sweat Economy

Sebastian Perez worked alone: The account of Perez's life and death is the result of dozens of interviews with family and friends, many of whom spoke to me off the record because of their immigration status.

fifteen million people: Jen M. Cox-Ganser et al. "Occupations by Proximity and Indoor /Outdoor Work: Relevance to COVID-19 in All Workers and Black/Hispanic Workers." *American Journal of Preventative Medicine* 60, no. 5 (2021), 621–628. https://doi.org/10.1016/j.amepre.2020.12.016

Kenton Scott Krupp: Umair Irfan. "Extreme Heat is Killing American Workers." *Vox,* July 21, 2021. https://www.vox.com/22560815/heat-wave-worker-extreme-climate-change-osha-workplace-farm-restaurant

a roofer died: Jamie Goldberg. "Two Oregon Businesses Whose Workers Died During Heat Wave Fight State Fines." *The Oregonian,* May 6, 2002. https://www.oregonlive.com/business/2022/05/two-oregon-businesses-whose-workers-died-during-heat-wave-fight-state-fines.html

photographs on social media: Twitter post by Teamsters for a Democratic Union, August 2, 2022. https://twitter.com/JimCShields/status/1554827644230717441?s=2

0&t=I_998pFBIDu2EiowCl7z9g

"People are dropping": Livia Albeck-Ripka. "UPS Drivers Say 'Brutal' Heat Is Endangering Their Lives." *New York Times,* August 20, 2022. https://www.nytimes.com/2022/08/20/business/ups-postal-workers-heat-stroke-deaths.html

Qatar: Annie Kelly, Niamh McIntyre, and Pete Pattirson. "Revealed: Hundreds of Migrant Workers Dying of Heat Stress in Qatar Each Year." *The Guardian,* October 2, 2019. https://www.theguardian.com/global-development/2019/oct/02/revealed-hundreds-of-migrant-workers-dying-of-heat-stress-in-qatar-each-year See also "World Cup 2022: How Many Migrant Workers Have Died in Qatar?" *Reuters,* December 14, 2022. https://www.reuters.com/lifestyle/sports/world-cup-2022-how-many-migrant-workers-have-died-qatar-2022-11-24/

Ken Caldeira: Quoted in Andrew Freedman and Jason Samenow, "Humidity and Heat Extremes Are on the Verge of Exceeding Limits of Human Survivability, Study Finds," *Washington Post,* May 8, 2020. https://www.washingtonpost.com/weather/2020/05/08/hot-humid-extremes-unsurvivable-global-warming/

thirty-five times more likely: Diane M.

Gubernot. "Characterizing Occupational Heat-Related Mortality in the United States, 2000–2010: An Analysis Using the Census of Fatal Occupational Injuries Database." *American Journal of Independent Medicine* 58, no. 2 (2015), 203–211. https://doi.org/10.1002/ajim.22381

poll of 2,176 farmworkers: "Farm Workers and Advocates on Heat Wave Affecting Ag Workers and the Urgency for Citizenship." United Farm Worker Foundation press call, July 7, 2021. https://www.ufwfoundation.org/farm_workers_and_advocates_heat_wave_affecting_ag_workers_adds_urgency_to_citizenship_push

chronic kidney disease: Richard J. Johnson. "Chronic Kidney Disease of Unknown Cause in Agricultural Communities." *New England Journal of Medicine* 380 (2019), 1843–1852. https://doi.org/10.1056/NEJMra1813869

An editorial: Cecilia Sorensen, MD, and Ramon Garcia-Trabanino, MD. "A New Era of Climate Medicine — Addressing Heat-Triggered Renal Disease." *New England Journal of Medicine* 381 (2019), 693–696. https://doi.org/10.1056/NEJMp1907859

twenty thousand additional injuries: Christopher Flavelle. "Work Injuries Tied to Heat Are Vastly Undercounted, Study Finds." *New York Times,* July 17, 2021. https://www.nytimes.com/2021/07/15/climate/heat-injuries.html

$100 billion in 2020: "Extreme Heat: the Economic and Social Consequences for the United States." Report by Vivid Economics and Adrienne Arsht-Rockefeller Foundation Resilience Center, August 2021. Accessed October 2022. https://www.atlanticcouncil.org/wp-content/uploads/2021/08/Extreme-Heat-Report-2021.pdf

Dhaka: "Hot Cities, Chilled Economies." Report by Vivid Economics and Adrienne Arsht-Rockefeller Foundation Resilience Center, August 2022. Accessed October 2022. https://onebillionresilient.org/hot-cities-chilled-economies-dhaka/

It would get to be 120: Ibid, 52.

"dark skin and light skin": Nina Jablonski. *Skin: A Natural History.* (Los Angeles and Berkeley: University of California Press, 2013), 78.

one abolitionist recalled: Walter Johnson. *River of Dark Dreams: Slavery and Empire in the Cotton Kingdom* (Cambridge, MA: Harvard University Press, 2013), 173.

Samuel Cartwright: Johnson, *River of Dark Dreams,* 199–204.

"The practice of negroes": Alan Derickson. "A Widespread Superstition: The Purported Invulnerability of Workers of Color to Occupational Heat Stress." *American Journal of Public Health* 109, no. 10 (2019), 1329–1335. https://doi.org/10.2105/AJPH.2019.305246

"The labor requiring exposure": Ibid.

"protected by the very nature": Ibid.

"The skull of the negro": Ibid.

"[The negro's] head": Ibid.

"God has adapted him": Ibid.

One investigator observed: Ibid.

Stereotypes for thirty-six groups: Ibid.

"The Mexicans": Ibid.

"The South was not fitted properly": Ibid.

"is a hot-weather plant": Ibid.

Ernst was flagged: "Farm Where Worker Died Cited Earlier for Safety Violations." *Seattle Times,* July 2, 2021. https://www.seattletimes.com/seattle-news/farm-where-worker-died-earlier-cited-for-safety-violations/

the nursery was cited: Jamie Goldberg. "Marion County Farm Where Worker Died Was Previously Cited for Workplace Safety Violations." *The Oregonian,* July 1, 2001. https://www.oregonlive.com/business/2021/07/marion-county-farm

-where-worker-died-was-previously-cited -for-workplace-violations.html

Chapter 9 — Ice at the End of the World

joint research project: The International Thwaites Glacier Collaboration is a shared research initiative of the US National Science Foundation and the UK Natural Environment Research Council. https://thwaitesglacier.org/

the Doomsday Glacier: Jeff Goodell. "The Doomsday Glacier." *Rolling Stone,* May 9, 2017. https://www.rollingstone.com/politics/politics-features/the-doomsday-glacier-113792/

three miles thick: "Quick Facts." National Snow and Ice Data Center website. Accessed October 2022. https://nsidc.org/learn/parts-cryosphere/ice-sheets/ice-sheet-quick-facts

1.2 to 3.2 feet: IPCC. *Climate Change 2021: The Physical Science Basis. Contribution of Working Group I to the Sixth Assessment Report of the Intergovernmental Panel on Climate Change* (Cambridge: Cambridge University Press, 2021), 1216–1217.

a caveat: Ibid.

ten feet by 2100: Jeff Goodell. "Will We Miss Our Last Chance to Save the World from Climate Change?" *Rolling Stone,*

December 22, 2016. https://www.rollingstone.com/politics/politics-features/will-we-miss-our-last-chance-to-save-the-world-from-climate-change-109426/

twenty feet higher: Andrea Dutton et al. "Sea-level Rise Due to Polar Ice-sheet Mass Loss During Past Warm Periods." *Science* 349, no. 6244 (2015). https://doi.org/10.1126/science.aaa4019

Larsen B ice shelf: Sabrina Shankman. "Trillion-Ton, Delaware-Size Iceberg Breaks Off Antarctica's Larsen C Ice Shelf." *Inside Climate News,* July 12, 2017. https://insideclimatenews.org/news/12072017/antarctica-larsen-c-ice-shelf-breaks-giant-iceberg

satellite imagery: Eric Rignot et al. "Four Decades of Antarctic Ice Sheet Mass Balance from 1979–2017." *Proceedings of the National Academy of Sciences* 116, no. 4 (2019), 1095–1103. https://doi.org/10.1073/pnas.1812883116

In the late 1960s: "The Scientist Who Predicted Ice Sheet Collapse — 50 Years Ago." *Nature* 554 (2018), 5–6. https://doi.org/10.1038/d41586-018-01390-x

marine ice-sheet instability: John Mercer. "West Antarctic Ice Sheet and CO_2 Greenhouse Effect: a Threat of Disaster."

Nature 271 (1978), 321–325. https://doi.org/10.1038/271321a0
1978 paper: Ibid.
"a major disaster": Ibid.

Chapter 10 — The Mosquito Is My Vector

half the people: Timothy C. Winegard. *The Mosquito: A Human History of Our Deadliest Predator* (New York: Dutton, 2019), 2.
slave ships: Joshua Sokol. "The Worst Animal in the World." *The Atlantic*. August 20, 2020. https://www.theatlantic.com/health/archive/2020/08/how-aedes-aegypti-mosquito-took-over-world/615328/
Ka-dinga pepo: J. G. Rigau-Pérez. "The Early Use of Break-Bone Fever (Quebranta huesos, 1771) and Dengue (1801) in Spanish." *American Journal of Tropical Medicine and Hygiene* 59, no. 2 (1998), 272–274. https://doi.org/10.4269/ajtmh.1998.59.272
only nine countries: World Health Organization. "Dengue and Severe Dengue." WHO factsheet. https://www.who.int/news-room/fact-sheets/detail/dengue-and-severe-dengue
in a hundred countries: Ibid.
390 million people: Ibid.
five billion people: Simon Hales et al. "Potential Effect of Population and Climate Changes on Global Distribution of Dengue

Fever: an Empirical Model." *The Lancet* 360, no. 9336 (2002), 830–834. https://doi.org/10.1016/S0140-6736(02)09964-6

ten thousand people: Jane P. Messina et al. "The Current and Future Global Distribution and Population at Risk of Dengue." *Nature Microbiology* 4 (2019), 1508–1515. https://doi.org/10.1038/s41564-019-0476-8

worsened more than half: Camilo Mora et al. "Over Half of Known Human Pathogenic Diseases Can Be Aggravated by Climate Change." *Nature Climate Change* 12 (2022), 869–875. https://doi.org/10.1038/s41558-022-01426-1

"a pandemic era": David M. Morens and Anthony S. Fauci. "Emerging Pandemic Diseases: How We Got to COVID-19." *Cell* 182, no. 5 (2020), 1077–1092. https://doi.org/10.1016/j.cell.2020.08.021

620 million: World Health Organization. WHO Coronavirus (COVID-19) Dashboard. Accessed October 2022. https://covid19.who.int/

six and a half million: Ibid.

fifty million people: "1918 Pandemic (H1N1 virus)." Centers for Disease Control website. Accessed October 2022. https://www.cdc.gov/flu/pandemic-resources/1918-pandemic-h1n1.html

Vibrio vulnificus: Ali Raj et al. "Deadly

Bacteria Lurk in Coastal Waters. Climate Change Increases the Risks." Center for Public Integrity website, Oct. 20, 2020. https://publicintegrity.org/environment/hidden-epidemics/vibrio-deadly-bacteria-coastal-waters-climate-change-health/

aftermath of Hurricane Ian: Frances Stead Sellars and Sabrina Malhi. "In Florida, Flesh-Eating Bacteria Follow in Hurricane Ian's Wake." *Washington Post,* October 18, 2022. https://www.washingtonpost.com/health/2022/10/18/flesh-eating-bacteria-florida/

four thousand species: Craig Welch. "Half of All Species on the Move — and We're Feeling it." *National Geographic,* April 26, 2017. https://www.nationalgeographic.com/science/article/climate-change-species-migration-disease

"A wild exodus": Shah, *The Next Great Migration,* 7.

1.7 million: P. Daszak et al. "IPBES Workshop Report on Biodiversity and Pandemics of the Intergovernmental Platform on Biodiversity and Ecosystem Services." IPBES secretariat, Bonn, Germany. https://doi.org/10.5281/zenodo.4147317

"massive simulation": Colin Carlson et al. "Climate Change Increases Cross-Species Viral Transmission Risk." *Nature* 607

(2022), 555–562. https://doi.org/10.1038/s41586-022-04788-w

"harrowing" Ed Yong. "We Created the 'Pandemicene'." *The Atlantic,* April 28, 2022. https://www.theatlantic.com/science/archive/2022/04/how-climate-change-impacts-pandemics/629699/

the small town of Hendra: David Quammen. *Spillover: Animal Infections and the Next Human Pandemic* (New York: W. W. Norton & Company, 2012), 13–19.

wildlife market in Wuhan: Michael Worobey et al. "The Huanan Seafood Wholesale Market in Wuhan was the early epicenter of the COVID-19 pandemic." *Science* 377, no. 6609 (2022), 951–959. https://doi.org/10.1126/science.abp8715

thirty years of detective work: Elahe Izadi. "Tracing the Long, Convoluted History of the AIDS Epidemic." *Washington Post,* February 24, 2015. https://www.washingtonpost.com/news/to-your-health/wp/2015/02/24/tracing-the-long-convoluted-history-of-the-aids-epidemic/

initially isolated: Quammen, *Spillover,* 324.

three hundred human cases: Ibid.

248 Nipah virus cases: Ibid.

three thousand species: Emily S. Rueb. "Peril on Wings: 6 of America's Most

Dangerous Mosquitoes." *New York Times,* June 28, 2016. https://www.nytimes.com/2016/06/29/nyregion/mosquitoes-diseases-zika-virus.html

"uniquely anthropophilic": David M. Morens et al. "Eastern Equine Encephalitis Virus — Another Emergent Arbovirus in the United States." *New England Journal of Medicine* 381 (2019), 1989–1992. https://doi.org/10.1056/NEJMp1914328

Nepal: World Health Organization. "Dengue Fever — Nepal." WHO Disease Outbreak News. Accessed October 2022. https://www.who.int/emergencies/disease-outbreak-news/item/2022-DON412

six hundred thousand people: World Health Organization. "Malaria." WHO factsheet. Accessed October 2022. https://www.who.int/news-room/fact-sheets/detail/malaria

higher and cooler regions: Colin Carlson et al. "Climate Change Increases Cross-Species Viral Transmission Risk." *Nature* 607 (2022), 555–562. https://doi.org/10.1038/s41586-022-04788-w

an additional seventy-six million: Sadie Ryan et al. "Shifting Transmission Risk for Malaria in Africa with Climate Change: a Framework for Planning and Intervention."

Malaria Journal 19, no. 170 (2020). https://doi.org/10.1186/s12936-020-03224-6

genetically modified: Emily Waltz. "Biotech Firm Announces Results from First US Trial of Genetically Modified Mosquitoes." *Nature News,* April 18, 2022. https://www.nature.com/articles/d41586-022-01070-x

"potential access to billions of humans": David M. Morens et al. "Eastern Equine Encephalitis Virus — Another Emergent Arbovirus in the United States." *New England Journal of Medicine* 381 (2019), 1989–1992. https://doi.org/10.1056/NEJMp1914328

Hyalommas running after people: Dennis Bente. "Hyalomma Marginatum Chasing by Sirri Kar." Accessed October 2022. https://www.youtube.com/watch?v=R_kGHqNpOQM.

fatality rate: World Health Organization. "Crimean-Congo Haemorrhagic Fever." WHO factsheet. Accessed October 2022. https://www.who.int/health-topics/crimean-congo-haemorrhagic-fever

seven hundred CCHF cases: Personal communication with the author, December 2020.

twenty different pathogens: Ibid.

"unparalleled threat": Meghan O'Rourke. "Lyme Disease is Baffling, Even to

Experts." *The Atlantic,* September 2019. https://www.theatlantic.com/magazine/archive/2019/09/life-with-lyme/594736/

first reported in 2017: "What You Need to Know About Asian Longhorned Ticks — A New Tick in the United States." Centers for Disease Control and Prevention. Accessed October 2022. https://www.cdc.gov/ticks/longhorned-tick/index.html

reduced production in dairy cattle: Ben Beard et al. "Multistate Infestation with the Exotic Disease–Vector Tick *Haemaphysalis longicornis.*" *Morbidity and Mortality Weekly Report* 67 (2018), 1310–1313. https://dx.doi.org/10.15585/mmwr.mm6747a3 external icon

Chapter 11 — Cheap Cold Air

"Broadway had open trolleys": Arthur Miller. "Before Air-Conditioning." *The New Yorker,* June 14, 1998. https://www.newyorker.com/magazine/1998/06/22/before-air-conditioning

"miasmas": Salvatore Basile. *Cool: How Air-conditioning Changed Everything* (New York: Fordham University Press, 2014), 23.

"Gorrie experimented": Ibid, 24.

"John Gorrie had actually produced": Ibid.

"Let the houses of warm countries": Eric Dean Wilson. *After Cooling: On Freon, Global Warming, and the Terrible Cost of Comfort* (New York: Simon & Schuster, 2021), 38.

"condition" the air: Ibid, 54.

"A fellow can make ten million": Thomas Thompson. *Blood and Money: A True Story of Murder, Passion, and Power* (New York: Doubleday, 1976), 19.

the FHA began including: Wilson, *After Cooling,* 161.

30 percent of homes: Ibid, 175.

Only 10 percent: Ibid, 176.

"The first fact of the day": William Styron. "As He Lay Dead, a Bitter Grief." *Life,* July 20, 1962.

"there are no seasons": William Faulkner. *The Reivers* (New York: Random House, 1962), 182.

President Franklin Delano Roosevelt: Joel D. Treese. "Keeping Cool in the White House." White House Historical Association website. Accessed October 2022. https://www.whitehousehistory.org/keeping-cool-in-the-white-house

President Lyndon Johnson: Ibid.

Aretha Franklin: Michael Simon. "Why Aretha Franklin Didn't Rehearse for Her VH1 'Divas Live' Performance." *Hollywood*

Reporter, August 21, 2018. https://www.hollywoodreporter.com/tv/tv-news/why-aretha-franklin-didnt-rehearse-her-vh1-divas-live-performance-1136286/

Between 1940 and 1980: Steven Johnson. *How We Got to Now: Six Inventions That Made the Modern World* (New York: Riverhead Books, 2014), 88.

Richard Nixon: Angie Maxwell. "What We Get Wrong About the Southern Strategy." *Washington Post,* July 26, 2019. https://www.washingtonpost.com/outlook/2019/07/26/what-we-get-wrong-about-southern-strategy/

fifteen thousand times: "Control of HFC-23 Emissions." EPA website. Accessed October 2022. https://www.epa.gov/climate-hfcs-reduction/control-hfc-23-emissions

20 percent: International Energy Agency. *The Future of Cooling* (Paris: IEA, 2019), 13. https://www.iea.org/reports/the-future-of-cooling

$3.7 billion: Yoshiyuki Osada. "Daikin Buys Goodman for $3.7 Billion, Gains North America Reach." *Reuters,* August 29, 2012. https://www.reuters.com/article/us-goodman-daikin/daikin-buys-goodman-for-3-7-billion-gains-north-america-reach-idUSBRE87S0A820120829

Comfortplex: HVAC Distributors.

“Daikin’s New Comfortplex Texas Facility Closer to Completion.” *HVAC News,* February 11, 2016. https://hvacdist.com/daikins-new-comfortplex-texas-facility-closer-to-completion/

one billion single-room: Sneha Sachar, Iain Campbell, and Ankit Kalanki. “Solving the Global Cooling Challenge: How to Counter the Climate Threat from Room Air Conditioners.” Rocky Mountain Institute report, 2018. Accessed October 2022. https://www.rmi.org/insight/solving_the_global_cooling_challenge

4.5 billion units: Ibid, 6.

“[In 2018] in Beijing”: Quoted in Stephen Buranyi. “The Air-conditioning Trap: How Cold Air Is Heating the World.” *The Guardian,* August 29, 2019. https://www.theguardian.com/environment/2019/aug/29/the-air-conditioning-trap-how-cold-air-is-heating-the-world

“It was markedly hotter”: State of Florida Administrative Hearing. “State of Florida, Agency of Health Care Administration vs. Rehabilitation Center at Hollywood Hills, LLC.” Case 17-005769, filed July 16, 2018, 16.

“Comfort is valued”: Daniel A. Barber. “After Comfort.” *Log* 47, 45–50.

“Comfort is destroying”: Ibid.

Chapter 12 — What You Can't See Won't Hurt You

hottest cities on Earth: Aryn Baker. "What It's Like Living in One of the Hottest Cities on Earth — Where It May Soon Be Uninhabitable." *Time,* September 12, 2019. https://time.com/longform/jacobabad-extreme-heat/

fewer than a million: Zofeen Ebrahim. "How Will Pakistan Stay Cool While Keeping Emissions in Check?" *The Third Pole,* March 11, 2022. https://www.thethirdpole.net/en/climate/pakistan-cooling-action-plan/

one-fifteenth: "CO_2 Emissions Per Capita." Worldometer website. Accessed October 2022. https://www.worldometers.info/co2-emissions/co2-emissions-per-capita/

Blue Marble: "Blue Marble — The Image of Earth from Apollo 17." NASA website. Accessed October 2022. https://www.nasa.gov/content/blue-marble-image-of-the-earth-from-apollo-17

Kent State: Bob Dyer. "Iconic Image from Kent State Shootings Stokes the Fires of Anti-Vietnam War Sentiment." *Akron Beacon Journal,* May 4, 2020. https://www.cincinnati.com/in-depth/news/history/2020/05/01/kent-state-shooting-photos-mary-ann-vecchio-impacts-nation-jeffrey-miller-john-filo/3055009001/

Falling Man: Tom Junod. "Who Was the Falling Man?" *Esquire,* September 9, 2021. https://www.esquire.com/news-politics/a48031/the-falling-man-tom-junod/

Selma-to-Montgomery: "From Selma to Montgomery: Stephen Somerstein's Photographs of the 1965 Civil Rights March," New York Historical Society & Library website. Accessed October 2022. https://www.nyhistory.org/blogs/from-selma-to-montgomery-stephen-somersteins-photographs-of-the-1965-civil-rights-march

Sahiwal: Baker, "What It's Like Living in One of the Hottest Cities."

Bashir's photo: Photo is posted on author's website, www.jeffgoodellwriter.com.

high temperature in Austin: Christopher Adams. "Austin Just Experienced Its Hottest Seven Day Stretch in History." KXAN weather blog. Accessed July 2022. https://www.kxan.com/weather/weather-blog/austin-just-experienced-its-hottest-7-day-stretch-in-history/

***New Yorker* cartoon:** Tristan Crocker. "Daily Cartoon: Thursday, May 12." *The New Yorker* website. Accessed October 2022. https://www.newyorker.com/cartoons/daily-cartoon/thursday-may-12th-dragon-warming

humidity in East Africa: Michael G. Just et al. "Human Indoor Climate Preferences Approximate Specific Geographies." *Royal Society Open Science* vol. 6, issue 3 (2019). https://doi.org/10.1098/rsos.180695

"harks back a hundred thousand years": Quoted in Nala Rogers, "Americans Make Their Homes Feel Like the African Savannah Where Humans First Evolved." *Inside Science,* March 19, 2019. https://www.insidescience.org/news/americans-make-their-homes-feel-african-savannah-where-humans-first-evolved

RealFeel: "What is the AccuWeather RealFeel Temperature?" *AcccuWeather News,* June 18, 2014. https://www.accuweather.com/en/weather-news/what-is-the-accuweather-realfeel-temperature/156655#

above thirty-eight miles per hour: NOAA, "Why Do We Name Tropical Storms and Hurricanes?" NOAA website. Accessed October 2022. https://oceanservice.noaa.gov/facts/storm-names.html

Charles Richter: "Richter Scale." US Geological Survey website. Accessed October 2022. https://earthquake.usgs.gov/learn/glossary/?term=richter%20scale

three categories: "Heat Watch vs. Warning." National Weather Service website.

Accessed October 2022. https://www.weather.gov/safety/heat-ww

Excessive Heat Warning: "WFO Non-Precipitation Weather Products (NPW) Specification." National Weather Service Instruction 10-515, December 27, 2019. Accessed October 2022. https://www.nws.noaa.gov/directives/sym/pd01005015curr.pdf

One 2018 study: David Hondula et al. "Spatial Analysis of United States National Weather Service Excessive Heat Warnings and Heat Advisories." *Bulletin of the American Meteorological Society* 103, no. 9 (2022) E2017–E2031. https://doi.org/10.1175/BAMS-D-21-0069.1

"the study": Ibid.

"a radical experiment": Catrin Einhorn and Christopher Flavelle. "A Race Against Time to Rescue a Reef Against Climate Change." *New York Times,* December 5, 2020. https://www.nytimes.com/2020/12/05/climate/Mexico-reef-climate-change.html

Solano's Storm: John Sledge. "Solano's Storm." *Mobile Bay,* September 3, 2020. https://mobilebaymag.com/solanos-storm/

"Great Miami Hurricane": Eric Jay Dolin, *A Furious Sky: the Five Hundred Year History of America's Hurricanes* (New York: Liveright, 2020), 123.

naming hurricanes: Ibid, 207.
gendered and misogynistic: Ibid, 210.
"himicanes": Ibid.
"In Mississippi": "Names for Heat Waves," *Northside Sun,* November 5, 2021. https://www.northsidesun.com/editorials-local-news-opinion/editorial-names-heatwaves
Lucifer: "Europe Swelters Under a Heat Wave Called 'Lucifer.'" *New York Times,* August 6, 2017. https://www.nytimes.com/2017/08/06/world/europe/europe-heat-wave.html
"Blisterer, Scorcher": Barbara Marshall. "European Heat Wave Called Lucifer: What Should We Call South Florida's?" *Palm Beach Post,* August 8, 2017. https://www.palmbeachpost.com/story/lifestyle/2017/08/08/european-heat-wave-called-lucifer/6763660007/
"Naming heat waves": Jason Samenow. "Heat Wave 'Hugo?' New Coalition Seeks to Name Hot Weather Like Hurricanes." *Washington Post,* August 6, 2020. https://www.washingtonpost.com/weather/2020/08/06/naming-heat-waves/
Forty-two heat researchers: Letter to Baughman McLeod and members of Adrienne Arsht-Rockefeller Foundation Resilience Center, August 17, 2020.
spatial synoptic classification: Laurence

Kalkstein et al. "A New Spatial Synoptic Classification: Application To Air-Mass Analysis." *International Journal of Climatology* 16, no. 9 (1996), 983–1004. https://doi.org/10.1002/(SICI)1097-0088(199609)16:9<983:AID-JOC61>3.0.CO;2-N

Seville committed: Ashifa Kassam. "Seville to Name and Classify Heat Waves in Effort to Protect Public." *The Guardian,* June 26, 2022. https://www.theguardian.com/world/2022/jun/26/seville-name-classify-heatwaves-effort-to-protect-public

earliest heat waves: Marco Trujillo. "Spain Melts Under the Earliest Heat Wave in Over 40 Years." *Reuters,* June 14, 2022. https://www.reuters.com/world/europe/spain-melts-under-earliest-heat-wave-over-40-years-2022-06-13/

swifts: Ashifa Kassam. " 'They're Being Cooked': Baby Swifts Die Leaving Nests as Heatwave Hits Spain." *The Guardian,* June 16, 2022. https://www.theguardian.com/world/2022/jun/16/spain-heatwave-baby-swifts-die-leaving-nest

a hundred chicks: Quoted in Ibid.

Heat Wave Zoe: Ciara Nugent. "Zoe, the World's First Named Heat Wave, Arrives in Seville." *Time,* July 25, 2022. https://time.com/6200153/first-named-heat-wave-zoe-seville/

a survey Arsht-Rock commissioned: Unpublished research shared with the author, October 2022.

embassies started tweeting: Akshaya Jha and Andrea La Nauze. "US Embassy Air-Quality Tweets Led to Global Health Benefits." *Proceedings of the National Academy of Sciences* 199, no. 44 (2022), e2201092119. https://doi.org/10.1073/pnas.2201092119

WMO issued a technical brief: World Meteorological Organization. "Considerations Regarding the Naming of Heat Waves." October 2022. https://library.wmo.int/index.php?lvl=notice_display&id=22190

health-based ranking system: Ashley R. Williams. "California Becomes First US State to Begin Ranking Extreme Heat Wave Events" *USA Today,* September 12, 2022. https://www.usatoday.com/story/news/nation/2022/09/12/california-becomes-first-state-start-ranking-extreme-heat-waves/8061975001/

Chapter 13 — Roast, Flee, or Act

full scope of the tragedy: An excellent, thorough account of the 2003 heat wave can be found in Richard Keller's *Fatal Isolation: The Devastating Heat Wave* of 2003 (Chicago: University of Chicago Press, 2015).

"a massive engorgement": Keller, *Fatal Isolation,* 41.
"One truck": Ibid.
fifteen thousand people: "Report on Behalf of the Commission of Inquiry on the Health and Social Consequences of the Heat Weather." French National Assembly, No. 1455, Vol. 1. https://www.assemblee-nationale.fr/12/rap-enq/r1455-t1.asp#P201_9399
"a pool of dried blood": Keller, *Fatal Isolation,* 41.
"In the blue hour": Alexandra Schwartz. "Paris Reborn and Destroyed." *The New Yorker,* March 19, 2014. https://www.newyorker.com/culture/culture-desk/paris-reborn-and-destroyed
final day of the climate summit: Jeff Goodell. "Will the Paris Climate Deal Save the World?" *Rolling Stone,* January 13, 2016. https://www.rollingstone.com/culture/culture-news/will-the-paris-climate-deal-save-the-world-56071/
entire New York City: Helene Chartier. "The Built Environment Industry Has a Huge Responsibility in the Climate Crisis." *ArchDaily,* September 22, 2022. https://www.archdaily.com/989430/the-built-environment-industry-has-a-huge-responsibility-in-the-climate-crisis

million trees: Dana Rubenstein. "A Million More Trees for New York City: Leaders Want a Greener Canopy." *New York Times,* February 12, 2022. https://www.nytimes.com/2022/02/12/nyregion/trees-parks-nyc.html

Seville: Laura Millan Lombrana. "One of Europe's Hottest Cities Is Using 1,000-Year-Old Technology to Combat Climate Change." *Bloomberg News,* August 18, 2022. https://www.bloomberg.com/news/articles/2022-08-18/one-of-europe-s-hottest-cities-has-a-climate-change-battle-plan

Freetown: Peter Yeung. "Africa's First Heat Officer Faces a Daunting Task." *Bloomberg News,* January 21, 2022. https://www.bloomberg.com/news/features/2022-01-21/how-africa-s-first-heat-officer-confronts-climate-change

Los Angeles: Christina Capatides. "Los Angeles Is Painting Some of Its Streets White and the Reasons Why Are Pretty Cool." *CBS News,* April 9 2018. https://www.cbsnews.com/news/los-angeles-is-painting-some-of-its-streets-white-and-the-reasons-why-are-pretty-cool/

India: "Green Roofs are Sprouting in India." *Times of India* website video. April 26, 2018. Accessed October 2022. https://times

ofindia.indiatimes.com/videos/city/chennai/green-roofs-are-sprouting-in-chennai/videoshow/63929209.cms?from=mdr

influential book: Franck Lirzin. *Paris face au changement climatique* (Paris: l'Aube, 2022).

"worn, wrecked and exhausted city": Alistair Horne. *Seven Ages of Paris* (New York: Vintage Books, 2004), 259.

major cholera epidemic: Ibid, 342.

Haussmann: Ibid, 363–386.

"a demolition artist": Ibid, 363.

four hundred thousand trees: Henry W. Lawrence. *City Trees: A Historical Geography from the Renaissance through the Nineteenth Century* (Charlottesville: University of Virginia Press, 2006), 237.

"some American Babylon": Joan DeJean. *How Paris Became Paris: The Invention of the Modern City* (New York: Bloomsbury, 2014), 300.

Émile Zola: Horne, *Seven Ages of Paris,* 374.

nearly 80 percent of the buildings: "Les Toits de Paris." Report from Atelier Parisian D'Urbanisme, October 2022, 14. Accessed November 2022. https://www.apur.org/sites/default/files/bd_toitures_paris.pdf

a hundred thousand buildings: Ibid.

UNESCO World Heritage: David Chazan. "Paris Wants Its 'Unique' Rooftops to Be Made UNESCO World Heritage Site." *The Telegraph,* January 28, 2015. https://www.telegraph.co.uk/news/world news/europe/france/11375145/Paris-wants -its-unique-rooftops-to-be-made-Unesco -world-heritage-site.html

194 degrees: Personal communication with Jacque Frier, December 2022.

white roofs: Nadia Razzhigaeva. "Are Cool Roofs the Future for Australian Cities?" UNSW Sydney news release, June 12, 2022. https://newsroom.unsw .edu.au/news/art-architecture-design /are-cool-roofs-future-australian-cities

fifteenth-century palazzos: Goodell, *The Water Will Come,* 116–144.

two miles of roadway: Lauren Ro and Alissa Walker. "Paris's Plan to Ban Cars from the Seine Holds up in Court." *Curbed,* October 25, 2018. https://archive.curbed .com/2016/9/27/13080078/paris-bans -cars-seine-right-bank-air-pollution -mayor-anne-hidalgo

250 miles of bike lanes: Madeline Schwartz. "Bike Lane to the Élysée" *New York Review,* March 24, 2022. https://www.nybooks .com/articles/2022/03/24/bike-lane-to-the -elysee-une-femme-francaise-hidalgo/

"My job is to transform": Kim Willsher. "Anne Hidalgo: 'Being Paris Mayor Is Like Piloting a Catamaran in a Gale.'" *The Guardian,* March 3, 2020. https://www.theguardian.com/world/2020/mar/03/anne-hidalgo-paris-mayor-second-term-interview

only 9 percent: Treepedia website. Accessed October 2022.

108.7 degrees: Anthony Cuthbertson. "Europe Heatwave: Paris Records its Hottest Temperature in History." *The Independent,* July 25, 2019. https://www.independent.co.uk/news/world/europe/paris-hottest-temperature-record-europe-heatwave-france-latest-a9019716.html

170,000 new trees: Vivian Song. "Admiring the Trees of Paris." *New York Times,* August 9, 2022. https://www.nytimes.com/2022/08/09/travel/paris-trees.html

300,000 trees: Marcus Fairs. "Forestami Project Will See 'One Tree for Every Inhabitant' Planted in Milan." *Dazeen,* September 24, 2021. https://www.dezeen.com/2021/09/24/forestami-project-trees-planted-milan/

three trillion trees: Elizabeth Pennisi. "Earth Home to 3 Trillion Trees, Half as Many as When Human Civilization Arose." *Science,* September 2, 2015.

https://www.science.org/content/article/earth-home-3-trillion-trees-half-many-when-human-civilization-arose

net loss each year: Ibid.

133 degrees: Manuel Ausloos. "Heatwave in Paris Exposes City's Lack of Trees." *Reuters,* August 4, 2022. https://www.reuters.com/world/europe/heatwave-paris-exposes-citys-lack-trees-2022-08-04/

$4,351.12 to plant a tree: Katherine McNenney. "It Costs $4,351.12 to Plant One Tree in LA." City Watch, December 1, 2022. https://www.citywatchla.com/index.php/neighborhood-politics/26031-it-costs-4-351-12-to-plant-one-tree-in-la

12 to 25 percent: "Tree and Shade Master Plan." City of Phoenix, 2010. Accessed November 2022. https://www.phoenix.gov/parkssite/Documents/PKS_Forestry/PKS_Forestry_Tree_and_Shade_Master_Plan.pdf

mulberry trees: Nick Kampouris. "Athens' Unique Mulberry Trees on the Brink of Extinction Due to Insect Damage." *Greek Reporter,* February 11, 2020. https://greekreporter.com/2020/02/11/athens-unique-mulberry-trees-on-the-brink-of-extinction-due-to-insect-damage/

Chicago and Milwaukee: Patty Wetli. "The Ash Trees Last Stand, and Why

It Matters" *WTTW News,* March 10, 2020. https://news.wttw.com/2020/03/04/ash-tree-last-stand-chicago-and-why-it-matters

1.4 million trees: Elizabeth Gamillo. "1.4 Million Trees May Fall to Invasive Insects by 2050." *Smithsonian,* March 17, 2022. https://www.smithsonianmag.com/smart-news/14-million-urban-trees-may-fall-to-invasive-insects-by-2050-180979752/

poisoned in 1989: Rosie Ninesling. "The Year a Love-Sick Occultist Poisoned Austin's Treaty Oak." *Austin Monthly,* March 2022. https://www.austinmonthly.com/the-year-a-love-sick-occultist-poisoned-austins-treaty-oak/

"Which Plant Where": Macquarie University website. Accessed November 2022. https://www.whichplantwhere.com.au/

Fallen superhero: Zach Hope. "Amid the Mourning, New Life for One of Melbourne's Most-Loved Trees." *The Age,* June 27, 2020. https://www.theage.com.au/national/victoria/amid-the-mourning-new-life-for-one-of-melbourne-s-most-loved-trees-20200619-p5545i.html

three-quarters of urban trees: "Texas Tree Death Toll from Drought Hits 5.6 Million, says Forest Service." *Houston Chronicle,*

February 17, 2012. https://www.chron.com/neighborhood/champions-klein/news/article/Texas-tree-death-toll-from-drought-hits-5-6-9505128.php

likely to die: Manuel Esperon-Rodriguez et al. "Climate Change Increases Global Risk to Urban Forests." *Nature Climate Change* 12 (2022), 950–955. https://doi.org/10.1038/s41558-022-01465-8

the most unequal tree cover: American Forests Tree Equity Score. American Forests website. Accessed October 2022. https://www.americanforests.org/tools-research-reports-and-guides/tree-equity-score/

Cheonggyecheon Stream: Colin Marshall. "Story of Cities #50: The Reclaimed Stream Bringing Life to the Heart of Seoul." *The Guardian,* May 25, 2016. https://www.theguardian.com/cities/2016/may/25/story-cities-reclaimed-stream-heart-seoul-cheonggyecheon

greenest city on Earth: Brian Barth. "Curitiba: the Greenest city on Earth." *The Ecologist,* March 15, 2004. https://theecologist.org/2014/mar/15/curitiba-greenest-city-earth

"We designed room for nature": Personal communication with the author, July 2022.

"Gucci biodiversity": Jared Green. "Earth Day Interview with Richard Weller: A Hopeful Vision for Global Conservation." *The Dirt,* April 18, 2022. https://dirt.asla.org/2022/04/18/earth-day-interview-with-richard-weller-a-bold-vision-for-global-conservation/

" 'zero milestones' ": Champs-Élysées Committee. "The Champs-Élysées — History & Perspectives." Exhibition at Pavillon de l'Arsenal, February 14 to September 13, 2020. https://www.pavillon-arsenal.com/fr/expositions/11463-champs-elysees.html

$300-million plan: Nadja Sayej. "Paris's Champs-Élysées Is Getting a Major Makeover — But What Does That Mean for the Locals?" *Architectural Digest,* January 29, 2021. https://www.architecturaldigest.com/story/pariss-champs-elysees-getting-major-makeover-what-does-that-mean-locals

"an extraordinary garden": Kim Willsher. "Paris Agrees to Turn Champs-Élysées into 'Extraordinary Garden.' " *The Guardian,* January 10, 2021. https://www.theguardian.com/world/2021/jan/10/paris-approves-plan-to-turn-champs-elysees-into-extraordinary-garden-anne-hidalgo

Lacaton & Vassal: Tom Ravenscroft. "Anne Lacaton and Jean-Philippe Vassal

Win Pritzker Architecture Prize 2021." *Dazeen,* March 16, 2021. https://www.dezeen.com/2021/03/16/anne-lacaton-jean-philippe-vassal-pritzker-architecture-prize-2021/

expand the Métro line: Shefali Anand. "With Grand Paris Express, Paris Hopes to Expand Its Borders — And Metropolitan Might." *Wall Street Journal,* July 2, 2018. https://www.wsj.com/articles/with-grand-paris-express-paris-hopes-to-expand-its-borders-and-metropolitan-might-1530553113

"Paris at 50°C": Emeline Cazi and Audrey Garric. "How Paris Is Preparing for Life 'at 50°C.'" *Le Monde,* July 17, 2022. https://www.lemonde.fr/en/environment/article/2022/07/17/how-paris-is-preparing-for-life-under-50-c_5990462_114.html

Chapter 14 — The White Bear

"to melt all the ice": Roni Dengler. "Ancient Mosses Suggest Canada's Baffin Island is the Hottest It's Been in 45,000 Years." *Science News,* October 30, 2017. https://www.science.org/content/article/ancient-mosses-suggest-canada-s-baffin-island-hottest-it-s-been-45000-years

two million times: Ashifa Kassam. "'Soul-crushing' Video of Starving Polar Bear

Exposes Climate Crisis, Experts Say." *The Guardian,* December 8, 2017. https://www.theguardian.com/environment/2017/dec/08/starving-polar-bear-arctic-climate-change-video

"rips the heart out": Matt Stevens. "Video of Starving Polar Bear 'Rips Your Heart Out of Your Chest.'" *New York Times,* December 11. 2017. https://www.nytimes.com/2017/12/11/world/canada/starving-polar-bear.html

"We care about the polar bears": Quoted in Katharine Hayhoe, "Yeah, the Weather Has Been Weird." *Foreign Policy,* May 31, 2017. https://foreignpolicy.com/2017/05/31/everyone-believes-in-global-warming-they-just-dont-realize-it/

"neither sail nor walk": Katheryn Schultz. "Literature's Arctic Obsession." *The New Yorker,* April 17, 2017. https://www.newyorker.com/magazine/2017/04/24/literatures-arctic-obsession

"a time when Aristotle": Ibid.

"farthest north": Ibid.

"sweet, and honorable, and sublime": Herman Melville. *Moby Dick: or, the Whale* (Berkeley: University of California Press, 1983), 196.

"irresponsible ferociousness": Ibid, 197.

"four times as fast": Mika Rantanen et

al. "The Arctic Has Warmed Nearly Four Times Faster than the Globe Since 1979." *Communications Earth & Environment* 3, no. 168 (2022). https://doi.org/10.1038/s43247-022-00498-3

Sir Wally Herbert: Wally Herbert and Roy M. Koerner. "The First Surface Crossing of the Arctic Ocean." *Geographical Journal* 136, no. 4 (1970), 511–533. https://doi.org/10.2307/1796181

twenty-five times more potent: EPA "Importance of Methane." Global Methane Initiative website. Accessed October 2022. https://www.epa.gov/gmi/importance-methane

reporting a story: Jeff Goodell. "Can Geoengineering Save the World?" *Rolling Stone,* October 4, 2011. https://www.rollingstone.com/politics/politics-news/can-geoengineering-save-the-world-238326/

fifteen million tons of sulfur dioxide: NASA "2018's Biggest Volcanic Eruption of Sulfur Dioxide." NASA website. Accessed October 2022. https://www.nasa.gov/feature/goddard/2019/2018-s-biggest-volcanic-eruption-of-sulfur-dioxide

ten million people: David Wallace-Wells. "Air Pollution Kills Ten Million People a Year. Why Do We Accept That As Normal?"

New York Times, July 8, 2022. https://www.nytimes.com/2022/07/08/opinion/environment/air-pollution-deaths-climate-change.html

ten to a hundred times larger: David Keith. "What's the Least Bad Way to Cool the Planet?" *New York Times,* October 1, 2022. https://www.nytimes.com/2021/10/01/opinion/climate-change-geoengineering.html

thirty-six billion tons: International Energy Agency. "Global CO_2 Emissions Rebounded to Their Highest Level in History in 2021." IEA press release, March 8, 2022. https://www.iea.org/news/global-co2-emissions-rebounded-to-their-highest-level-in-history-in-2021

"Male bears, especially subadults": Personal communication with the author, July 2021.

Doug Peacock: Doug Peacock. *Grizzly Years: In Search of the American Wilderness* (New York: Holt Paperbacks, 1996), 139.

Epilogue — Beyond Goldilocks

President Lyndon B. Johnson: Dana Nuccitelli. "Scientists Warned the US President About Global Warming 50 Years Ago Today." *The Guardian,* November 15, 2015. https://www.theguardian.com

/environment/climate-consensus-97-percent/2015/nov/05/scientists-warned-the-president-about-global-warming-50-years-ago-today

ExxonMobil: Neela Banerjee, Lisa Song, and David Hasemyer. "Exxon's Own Research Confirmed Fossil Fuels' Role in Global Warming Decades Ago." *Inside Climate News,* September 16, 2015. https://insideclimatenews.org/news/16092015/exxons-own-research-confirmed-fossil-fuels-role-in-global-warming/

a thousand deaths a day: Holly Yan and Christina Maxouris. "The US Just Topped 1,100 Coronavirus Deaths a Day. One State Is Getting National Guard Help, and Others Keep Breaking Records." *CNN,* October 22, 2020. https://www.cnn.com/2020/10/22/health/us-coronavirus-thursday/index.html

43,000 deaths: "Newly Released Estimates Show Traffic Fatalities Reached a 16-Year High in 2021." National Highway Traffic Safety Administration press release, May 17, 2022. https://www.nhtsa.gov/press-releases/early-estimate-2021-traffic-fatalities

"We are confronted simultaneously": Brink Lindsey. "What Is the Permanent Problem?" Brink Lindsey website. September 28, 2022. https://brinklindsey.substack

.com/p/what-is-the-permanent-problem

"Humanity retains an enormous amount": David Wallace-Wells. "Beyond Catastrophe A New Climate Reality is Coming into View." *New York Times Magazine,* October 26, 2022. https://www.nytimes.com/interactive/2022/10/26/magazine/climate-change-warming-world.html

"the Venus Express": James Hansen. *Storms of My Grandchildren: The Truth About the Coming Climate Catastrophe and Our Last Chance to Save Humanity* (New York: Bloomsbury, 2009), 175.

"which evolutionary pathways": Elizabeth Kolbert. *The Sixth Extinction: An Unnatural History* (New York: Henry Holt, 2014), 268–269.

"How do we face the truth": "Dr. Ayana Elizabeth Johnson: Hope is Courage and Taking Action Together. *The Jane Goodall Hopecast,*" May 31, 2021. https://news.janegoodall.org/2021/06/01/jane-goodall-hopecast-podcast-ep-15-dr-ayana-elizabeth-johnson/

SELECTED BIBLIOGRAPHY

Anderson, Warwick. *The Cultivation of Whiteness: Science, Health, and Racial Destiny in Australia.* Durham, NC: Duke University Press, 2006.

Basile, Salvatore. *Cool: How Air-conditioning Changed Everything*. New York: Fordham University Press, 2014.

Brown, G. I. *Scientist, Soldier, Statesman, Spy: Count Rumford, the Extraordinary Life of a Scientific Genius.* United Kingdom: Sutton Publishing, 1999.

Brown, Sanborn C. *Benjamin Thompson, Count Rumford.* Cambridge, MA: MIT Press, 1981.

Bruchag, Joseph. *The Native American Sweat Lodge: History and Legends*. Freedom, CA: The Crossing Press, 1993.

Brusatte, Steve. *The Rise and Fall of the Dinosaurs: A New History of a Lost World.* New York: William Morrow, 2020.

Bryson, Bill. *The Body: A Guide for*

Occupants. New York: Doubleday, 2019.

Cardwell, D. S. L. *From Watt to Clausius: The Rise of Thermodynamics in the Early Industrial Age.* Ithaca, NY: Cornell University Press, 1971.

Christiansen, Rupert. *City of Light: The Making of Modern Paris*. New York: Basic Books, 2018.

Cunningham, Sophie. *City of Trees: Essays on Life, Death & the Need for a Forest.* Melbourne: Text Publishing, 2019.

Dartnell, Lewis. *Origins: How Earth's History Shaped Human History*. New York: Basic Books, 2019.

DeJean, Joan. *How Paris Became Paris: The Invention of the Modern City*. New York: Bloomsbury, 2014.

Dessler, Andrew. *Introduction to Modern Climate Change* (Third Edition). United Kingdom: Cambridge University Press, 2022.

Dolin, Eric Jay. *A Furious Sky: The Five Hundred Year History of America's Hurricanes*. New York: Liveright, 2020.

Dunar, Andrew J. and Dennis McBride. *Building the Hoover Dam: An Oral History of the Great Depression.* Reno: University of Nevada Press, 1993.

Engelhard, Michael. *Ice Bear: the Cultural History of an Arctic Icon.* Seattle: University of Washington Press, 2017.

Everts, Sarah. *The Joy of Sweat: The Strange Science of Perspiration*. New York: W. W. Norton & Company, 2021.
Farmer, Jared. *Elderflora: A Modern History of Ancient Trees.* New York: Basic Books, 2022.
Fiennes, Ranulph. *Heat: Extreme Adventures at the Highest Temperatures on Earth*. London: Simon & Schuster, 2015.
Gammage, Bill. *The Biggest Estate on Earth: How Aborigines Made Australia*. Crow's Nest, Australia: Allen & Unwin, 2011.
Gisolfi, Carl V. and Francisco Mora. *The Hot Brain: Survival, Temperature, and the Human Body.* Cambridge, MA: MIT Press, 2000.
Goldstein, Martin and Inge F. Goldstein. *The Refrigerator and the Universe: Understanding the Laws of Energy.* Cambridge, MA: Harvard University Press, 1993.
Greene, Brian. *Until the End of Time: Mind, Matter, and Our Search for Meaning in an Evolving Universe.* New York: Knopf, 2020.
Hansen, James. *Storms of My Grandchildren: The Truth about the Coming Climate Catastrophe and Our Last Chance to Save Humanity*. New York: Bloomsbury Press, 2009.
Horne, Alistair. *Seven Ages of Paris*. New York: Vintage Books, 2004.
Hunt, Kevin. *Chimpanzee: Lessons from*

Our Sister Species. Cambridge, England: Cambridge University Press, 2021.

Hutchinson, Alex. *Endure: The Mind, Body, and the Curiously Elastic Limits of Human Performance*. New York: William Morrow, 2018.

Jackson, Roland. *The Ascent of John Tyndall: Victorian Scientist, Mountaineer, and Public Intellectual*. Oxford, England: Oxford University Press, 2018.

Jablonski, Nina. *Skin: A Natural History*. Los Angeles and Berkeley: University of California Press, 2013.

Jacquart, Danielle and Claude Thomasset. *Sexuality and Medicine in the Middle Ages*. Princeton, NJ: Princeton University Press, 1988.

Johnson, Lizzie. *Paradise: One Town's Struggle to Survive an American Wildfire*. New York: Crown, 2021.

Johnson, Steven. *How We Got to Now: Six Inventions That Made the Modern World*. New York: Riverhead Books, 2014.

Johnson, Walter. *River of Dark Dreams: Slavery and Empire in the Cotton Kingdom*. Cambridge, MA: Harvard University Press, 2013.

Kamler, Kenneth. *Surviving the Extremes: What Happens to the Body and Mind at the Limits of Endurance*. New York: St. Martin's Press, 2003.

Keller, Richard. *Fatal Isolation: The Devastating Paris Heat Wave of 2003*. Chicago: University of Chicago Press, 2014.

Khanna, Parag. *Move: The Forces Uprooting Us*. New York: Scribner, 2001.

Klinenberg, Eric. *Heat Wave: A Social Autopsy of Disaster in Chicago*. Chicago: University of Chicago Press, 2002.

Kolbert, Elizabeth. *The Sixth Extinction: An Unnatural History*. New York: Henry Holt, 2014.

Kolbert, Elizabeth. *Under a White Sky: The Nature of the Future.* New York: Crown, 2021.

Lawrence, Henry W. *City Trees: A Historical Geography from the Renaissance through the Nineteenth Century.* Charlottesville: University of Virginia Press, 2006.

Lirzin, Franck. *Paris face au changement climatique*. Paris: l'Aube, 2022.

Lovelock, James with Bryan Appleyard. *Novacene: The Coming Age of Hyperintelligence.* London: Penguin Books, 2019.

Mann, Michael. *The New Climate War.* New York: Public Affairs, 2021.

Marsden, Ben. *Watt's Perfect Engine: Steam & the Age of Invention.* Duxford, England: Icon Books, 2004.

Marshall, S. L. A. *Battles in the Monsoon: Campaigning in the Central Highlands,*

Vietnam, Summer 1966. New York: William Morrow, 1967.

McDonald, Robert I. *Conservation for Cities: How to Plan and Build Natural Infrastructure.* Washington, DC: Island Press, 2015.

Merrill, Jane. *Sex and the Scientist: The Indecent Life of Benjamin Thompson, Count Rumford (1753–1814).* Jefferson, North Carolina: McFarland & Company, 2018.

Miller, Henry. *The Air-conditioned Nightmare.* New York: New Directions, 1945.

Monbiot, George. *Regenesis: Feeding the World Without Devouring the Planet.* New York: Penguin Books, 2022.

Nelson, Scott Reynolds. *Oceans of Grain: How American Wheat Remade the World.* New York: Basic Books, 2022.

Otto, Friedricke. *Angry Weather: Heat Waves, Floods, Storms and the New Science of Climate Change.* Berkeley/Vancouver: Greystone, 2020.

Quammen, David. *Breathless: The Scientific Race to Defeat a Deadly Virus.* New York: Simon & Schuster, 2022.

Quammen, David. *Spillover: Animal Infections and the Next Human Pandemic.* New York: W. W. Norton & Company, 2012.

Raff, Jennifer. *Origin: A Genetic History of the Americas.* New York: Twelve, 2022.

Saladino, Dan. *Eating to Extinction: The*

World's Rarest Foods and Why We Need to Save Them. New York: Farrar, Straus, and Giroux, 2021.
Sciubba, Jennifer D. *8 Billion and Counting: How Sex, Death, and Migration Shape the World*. New York: W. W. Norton & Company, 2022.
Shah, Sonia. *The Next Great Migration: The Beauty and Terror of Life on the Move*. New York: Bloomsbury, 2020.
Steffen, Alex. *Carbon Zero: Imagining Cities That Can Save the Planet*. Self-published, 2013.
Stirling, Ian. *Polar Bears: The Natural History of a Threatened Species*. Markham, Ontario: Fitzhenry & Whiteside, 2011.
Streever, Bill. *Heat: Adventures in the World's Fiery Places*. New York: Little, Brown, 2013.
Thompson, Thomas. *Blood and Money: A True Story of Murder, Passion, and Power*. New York: Doubleday, 1976.
Tyndall, John. *Heat: A Mode of Motion*. London: Spottiswoode and Co., 1875.
Uglow, Jenny. *The Lunar Men: Five Friends Whose Curiosity Changed the World*. New York: Farrar, Straus and Giroux, 2002.
Urrea, Luis Alberto. *The Devil's Highway: A True Story*. New York: Little, Brown, 2004.
Von Baeyer, Hans Christian. *Warmth*

Disperses and Time Passes: The History of Heat. New York: Modern Library, 1999.
Wallace-Wells, David. *The Uninhabitable Earth: Life After Warming.* New York: Tim Duggan Books, 2019.
Weart, Spencer R. *The Discovery of Global Warming.* Cambridge, MA: Harvard University Press, 2008.
Wilson, Ben. *Urban Jungle: The History and Future of Nature in the City.* New York: Doubleday, 2023.
Wilson, Eric Dean. *After Cooling: On Freon, Global Warming, and the Terrible Cost of Comfort.* New York: Simon & Schuster, 2021.
Winegard, Timothy C. *The Mosquito: A Human History of Our Deadliest Predator.* New York: Dutton, 2019.
Woollings, Tim. *Jetstream.* Oxford, England: Oxford University Press, 2020.
Zimmer, Carl. *Life's Edge: The Search for What It Means to Be Alive.* New York: Dutton, 2021.

ABOUT THE AUTHOR

Jeff Goodell is the author of seven books, including *The Water Will Come: Rising Seas, Sinking Cities, and the Remaking of the Civilized World,* which was a *New York Times* Critics Top Book of 2017. He has covered climate change for more than two decades at *Rolling Stone* and many other publications. He is a Senior Fellow at the Adrienne Arsht-Rockefeller Foundation Resilience Center and a 2020 Guggenheim Fellow.

The employees of Thorndike Press hope you have enjoyed this Large Print book. All our Thorndike Large Print titles are designed for easy reading, and all our books are made to last. Other Thorndike Press Large Print books are available at your library, through selected bookstores, or directly from us.

For information about titles, please call:

(800) 223-1244

or visit our website at:

http://gale.cengage.com/thorndike